Equitable and Engaging Mathematics Teaching

A Guide to Disrupting Hierarchies in the Classroom

Electronic ISBN 978-1-61444-333-9

Print ISBN 978-0-88385-212-5

Published in the United States of America

Equitable and Engaging Mathematics Teaching

A Guide to Disrupting Hierarchies in the Classroom

Daniel Reinholz

MAA

Published and Distributed by

The Mathematical Association of America

The MAA Notes Series, started in 1982, addresses a broad range of topics and themes of interest to all who are involved with undergraduate mathematics. The volumes in this series are readable, informative, and useful, and help the mathematical community keep up with developments of importance to mathematics.

MAA Notes

14. Mathematical Writing, by *Donald E. Knuth, Tracy Larrabee, and Paul M. Roberts.*
16. Using Writing to Teach Mathematics, *Andrew Sterrett,* Editor.
17. Priming the Calculus Pump: Innovations and Resources, Committee on Calculus Reform and the First Two Years, a subcommittee of the Committee on the Undergraduate Program in Mathematics, *Thomas W. Tucker,* Editor.
18. Models for Undergraduate Research in Mathematics, *Lester Senechal,* Editor.
19. Visualization in Teaching and Learning Mathematics, Committee on Computers in Mathematics Education, *Steve Cunningham and Walter S. Zimmermann,* Editors.
20. The Laboratory Approach to Teaching Calculus, *L. Carl Leinbach et al.,* Editors.
21. Perspectives on Contemporary Statistics, *David C. Hoaglin and David S. Moore,* Editors.
22. Heeding the Call for Change: Suggestions for Curricular Action, *Lynn A. Steen,* Editor.
24. Symbolic Computation in Undergraduate Mathematics Education, *Zaven A. Karian,* Editor.
25. The Concept of Function: Aspects of Epistemology and Pedagogy, *Guershon Harel and Ed Dubinsky,* Editors.
26. Statistics for the Twenty-First Century, *Florence and Sheldon Gordon,* Editors.
27. Resources for Calculus Collection, Volume 1: Learning by Discovery: A Lab Manual for Calculus, *Anita E. Solow,* Editor.
28. Resources for Calculus Collection, Volume 2: Calculus Problems for a New Century, *Robert Fraga,* Editor.
29. Resources for Calculus Collection, Volume 3: Applications of Calculus, *Philip Straffin,* Editor.
30. Resources for Calculus Collection, Volume 4: Problems for Student Investigation, *Michael B. Jackson and John R. Ramsay,* Editors.
31. Resources for Calculus Collection, Volume 5: Readings for Calculus, *Underwood Dudley,* Editor.
32. Essays in Humanistic Mathematics, *Alvin White,* Editor.
33. Research Issues in Undergraduate Mathematics Learning: Preliminary Analyses and Results, *James J. Kaput and Ed Dubinsky,* Editors.
34. In Eves' Circles, *Joby Milo Anthony,* Editor.
35. You're the Professor, What Next? Ideas and Resources for Preparing College Teachers, The Committee on Preparation for College Teaching, *Bettye Anne Case,* Editor.
36. Preparing for a New Calculus: Conference Proceedings, *Anita E. Solow,* Editor.

37. A Practical Guide to Cooperative Learning in Collegiate Mathematics, *Nancy L. Hagelgans, Barbara E. Reynolds, SDS, Keith Schwingendorf, Draga Vidakovic, Ed Dubinsky, Mazen Shahin, G. Joseph Wimbish, Jr.*
38. Models That Work: Case Studies in Effective Undergraduate Mathematics Programs, *Alan C. Tucker,* Editor.
39. Calculus: The Dynamics of Change, CUPM Subcommittee on Calculus Reform and the First Two Years, *A. Wayne Roberts,* Editor.
40. Vita Mathematica: Historical Research and Integration with Teaching, *Ronald Calinger,* Editor.
41. Geometry Turned On: Dynamic Software in Learning, Teaching, and Research, *James R. King and Doris Schattschneider,* Editors.
42. Resources for Teaching Linear Algebra, *David Carlson, Charles R. Johnson, David C. Lay, A. Duane Porter, Ann E. Watkins, William Watkins,* Editors.
43. Student Assessment in Calculus: A Report of the NSF Working Group on Assessment in Calculus, *Alan Schoenfeld,* Editor.
44. Readings in Cooperative Learning for Undergraduate Mathematics, *Ed Dubinsky, David Mathews, and Barbara E. Reynolds,* Editors.
45. Confronting the Core Curriculum: Considering Change in the Undergraduate Mathematics Major, *John A. Dossey,* Editor.
46. Women in Mathematics: Scaling the Heights, *Deborah Nolan,* Editor.
47. Exemplary Programs in Introductory College Mathematics: Innovative Programs Using Technology, *Susan Lenker,* Editor.
48. Writing in the Teaching and Learning of Mathematics, *John Meier and Thomas Rishel.*
49. Assessment Practices in Undergraduate Mathematics, *Bonnie Gold,* Editor.
50. Revolutions in Differential Equations: Exploring ODEs with Modern Technology, *Michael J. Kallaher,* Editor.
51. Using History to Teach Mathematics: An International Perspective, *Victor J. Katz,* Editor.
52. Teaching Statistics: Resources for Undergraduate Instructors, *Thomas L. Moore,* Editor.
53. Geometry at Work: Papers in Applied Geometry, *Catherine A. Gorini,* Editor.
54. Teaching First: A Guide for New Mathematicians, *Thomas W. Rishel.*
55. Cooperative Learning in Undergraduate Mathematics: Issues That Matter and Strategies That Work, *Elizabeth C. Rogers, Barbara E. Reynolds, Neil A. Davidson, and Anthony D. Thomas,* Editors.
56. Changing Calculus: A Report on Evaluation Efforts and National Impact from 1988 to 1998, *Susan L. Ganter.*
57. Learning to Teach and Teaching to Learn Mathematics: Resources for Professional Development, *Matthew Delong and Dale Winter.*
58. Fractals, Graphics, and Mathematics Education, Benoit Mandelbrot and Michael Frame, Editors.
59. Linear Algebra Gems: Assets for Undergraduate Mathematics, *David Carlson, Charles R. Johnson, David C. Lay, and A. Duane Porter,* Editors.
60. Innovations in Teaching Abstract Algebra, *Allen C. Hibbard and Ellen J. Maycock,* Editors.
61. Changing Core Mathematics, *Chris Arney and Donald Small,* Editors.
62. Achieving Quantitative Literacy: An Urgent Challenge for Higher Education, *Lynn Arthur Steen.*
64. Leading the Mathematical Sciences Department: A Resource for Chairs, *Tina H. Straley, Marcia P. Sward, and Jon W. Scott,* Editors.
65. Innovations in Teaching Statistics, *Joan B. Garfield,* Editor.
66. Mathematics in Service to the Community: Concepts and models for service-learning in the mathematical sciences, *Charles R. Hadlock,* Editor.
67. Innovative Approaches to Undergraduate Mathematics Courses Beyond Calculus, *Richard J. Maher,* Editor.
68. From Calculus to Computers: Using the last 200 years of mathematics history in the classroom, Amy Shell-Gellasch and Dick Jardine, Editors.
69. A Fresh Start for Collegiate Mathematics: Rethinking the Courses below Calculus, *Nancy Baxter Hastings,* Editor.
70. Current Practices in Quantitative Literacy, *Rick Gillman*, Editor.
71. War Stories from Applied Math: Undergraduate Consultancy Projects, *Robert Fraga*, Editor.
72. Hands On History: A Resource for Teaching Mathematics, *Amy Shell-Gellasch*, Editor.

73. Making the Connection: Research and Teaching in Undergraduate Mathematics Education, *Marilyn P. Carlson and Chris Rasmussen*, Editors.

74. Resources for Teaching Discrete Mathematics: Classroom Projects, History Modules, and Articles, *Brian Hopkins*, Editor.

75. The Moore Method: A Pathway to Learner-Centered Instruction, *Charles A. Coppin, W. Ted Mahavier, E. Lee May, and G. Edgar Parker*.

76. The Beauty of Fractals: Six Different Views, *Denny Gulick and Jon Scott*, Editors.

77. Mathematical Time Capsules: Historical Modules for the Mathematics Classroom, *Dick Jardine and Amy Shell-Gellasch*, Editors.

78. Recent Developments on Introducing a Historical Dimension in Mathematics Education, *Victor J. Katz and Costas Tzanakis*, Editors.

79. Teaching Mathematics with Classroom Voting: With and Without Clickers, *Kelly Cline and Holly Zullo*, Editors.

80. Resources for Preparing Middle School Mathematics Teachers, *Cheryl Beaver, Laurie Burton, Maria Fung, and Klay Kruczek*, Editors.

81. Undergraduate Mathematics for the Life Sciences: Models, Processes, and Directions, *Glenn Ledder, Jenna P. Carpenter, and Timothy D. Comar*, Editors.

82. Applications of Mathematics in Economics, *Warren Page*, Editor.

83. Doing the Scholarship of Teaching and Learning in Mathematics, *Jacqueline M. Dewar and Curtis D. Bennett*, Editors.

84. Insights and Recommendations from the MAA National Study of College Calculus, *David Bressoud, Vilma Mesa, and Chris Rasmussen,* Editors.

85. Beyond Lecture: Resources and Pedagogical Techniques for Enhancing the Teaching of Proof-Writing Across the Curriculum, *Rachel Schwell, Aliza Steurer and Jennifer F. Vasquez,* Editors.

86. Using the Philosophy of Mathematics in Teaching Undergraduate Mathematics, *Bonnie Gold, Carl E. Behrens, and Roger A. Simons,* Editors.

87. The Courses of History: Ideas for Developing a History of Mathematics Course, *Amy Shell-Gellasch and Dick Jardine,* Editors.

88. Shifting Contexts, Stable Core: Advancing Quantitative Literacy in Higher Education, *Luke Tunstall, Gizem Karaali, and Victor Piercey,* Editors.

89. MAA Instructional Practices Guide, *Martha L. Abell, Linda Braddy, Doug Ensley, Lewis Ludwig, Hortensia Soto,* Project Leadership Team.

90. What Could They Possibly Be Thinking!?! Understanding your college math students, *Dave Kung and Natasha Speer*.

91. Mathematical Themes in a First-Year Seminar, *Jennifer Schaefer, Jennifer Bowen, Mark Kozek, and Pamela Pierce,* Editors.

92. Addressing Challenges to the Precalculus to Calculus II Sequence through Case Studies: Report based on the National Science Foundation Funded Project Precalculus through Calculus II, *Estrella Johnson, Naneh Apkarian, Kristen Vroom, Antonio Martinez, Chris Rasmussen, and David Bressoud*, Editors.

93. Engaging Students in Introductory Mathematics Courses through Interdisciplinary Partnerships: The SUMMIT-P Model *Susan L. Ganter, Debra Bourdeau, Victor Piercey, and Afroditi V. Filippas,* Editors.

94. Expanding Undergraduate Research in Mathematics: Making UR more inclusive, *Michael Dorff, Jan Rychtář, Dewey Taylor,* Editors.

95. Sharing and Storing Knowledge about Teaching Undergraduate Mathematics: An Introduction to a Written Genre for Sharing Lesson-specific Instructional Knowledge, *Douglas Lyman Corey and Steven R. Jones*, Editors.

96. Justice Through the Lens of Calculus: Framing New Possibilities for Diversity, Equity, and Inclusion, *Matthew Voigt, Jess Ellis Hagman, Jessica Gehrtz, Brea Ratliff, Nathan Alexander, and Rachel Levy*, Editors.

97. Equitable and Engaging Mathematics Teaching: A Guide to Disrupting Hierarchies in the Classroom, *Daniel Reinholz*

MAA Service Center
P.O. Box 91112
Washington, DC 20090-1112
1-800-331-1MAA FAX: 1-301-206-9789

Contents

1

Introduction

1.1 Author Positionality

My background, education, and research perspectives have all shaped the book you are reading, so I want to tell you a bit more about myself so you can understand why the book looks like it does. This work is primarily situated in the US context, which has a particular legacy of indigenous genocide, enslavement, xenophobia, colonialism, patriarchy, militarism, global interventionism, and other forms of oppression. If our goal is to teach equitably in the US, then we simply cannot ignore this history. Equitable teaching requires a historical perspective. Moreover, while many ideas from this book could be applied to other contexts, they would need to be adapted to account for differences in culture, hierarchies of power, historical legacies and so forth.

As a white and masculine-presenting person in the US, I have grown up with unearned racial and gender privilege. I have never had to question whether or not I belong in a mathematics classroom based on my race and gender. In many ways, this provided me with access to study and receive advanced degrees in engineering (BS), mathematics (MS), and mathematics education (PhD). Given that the majority of mathematicians in the US are white men, my conviction is that we have a responsibility for addressing historical inequities in our discipline that still exist today, as collectively, we have played a predominant role in creating these inequities in the first place. Much of my professional career has been dedicated to working with other white educators so that we can become more aware of the ways that we (often unintentionally) perpetuate harm in our classrooms, and so that we can reduce that harm and make mathematics more equitable, inclusive, useful, and joyful for everyone. Thus, while I write this book for all mathematics educators, I especially write for those of us who operate along a number of axes of privilege, and who are new to using equitable teaching practices in their classrooms.

Despite my current levels of professional success, I am a first-generation college student and high school drop out. I am a parent of two young children, and the adventure of being with two beautiful young humans has profoundly shaped who I am. I am autistic, non-binary, disabled, and have been hospitalized more times than I can recall. These identities are largely invisible to those around me, except those who I know closely or those who know how to see them. I was an adult when I learned that I am autistic, and every day I am still discovering exactly what that means to me. Ironically, despite being stigmatized, my autistic brain is inclined towards mathematical structures. Given this background, and the challenges that can come with it, I have made an extra effort to include material on disability, which is not often talked about in our field.

In particular, my worldview is grounded in disability justice [124]. Disability justice contrasts a medicalized approach to disability, which fixates on individual flaws and accommodations. Instead, disability justice situates individual embodied experiences within broader social structures. Disability justice attends to the pernicious role of ableism in society, especially as it undergirds and amplifies other systems of oppression. A disability justice approach is inherently intersectional, collaborative, and liberatory, insofar that it focuses on dismantling *all* systems that create hierarchy rather than simply replacing old hierarchies with new ones. This framework was developed in activist spaces—not the academy—through the liberatory work of disabled people of color and disabled queer activists. Disability justice attends explicitly to access, wholeness, embodiment, and humanity in the quest for liberation. It aims to disrupt ideas of normal, normative, and perfect, recognizing the beauty of the imperfect [174]. Although this framework was created by disabled people, it speaks broadly to the needs and liberation of *all* people. Indeed, the very concepts of disabled and nondisabled are social constructions, which are grounded in specific social and historical contexts.

The person who I am is a complex intersection of this variety of identities and my own lived experiences. These identities shape my viewpoint and experience of the world, and they shape the book I have written. I don't claim to understand or speak for other intersections of identities (other than my own), and much of what I have learned over the years is from people who have different identities than I do. In this book I do my best to be aware of my own biases and the limitations of my own positioning, while authentically representing the scholarship of mathematics education and instructional change. I recognize the inherent limitations of a single author writing a book and have done my best to represent the wisdom and viewpoints of many colleagues and collaborators. I aim to be practical rather than definitive. This book compiles things that I have learned over the past decade and that I have found useful in supporting mathematics instructors to disrupt hierarchies in their classrooms. I hope you find them useful too. Much of this work has taken place in the US, and may not directly translate to other contexts.

As an instructor, I have taught in a wide variety of classroom contexts. I have taught mathematics to undergraduates at a research-intensive and primarily white institution. I have also taught in racially and socioeconomically diverse settings like community colleges and my current PhD-granting institution. I have worked with incarcerated people at San Quentin Prison. I have taught in-person and online. I have a K–12 teaching credential, which means I completed the entire program of study as well as field experiences and student teaching. I have extensive experience running professional learning for staff and faculty. I draw from these various teaching perspectives in this book. Although I have worked across a variety of settings, there are some contexts in which I have little experience (e.g., teaching in a residential college or cohort-based program).

The research work I draw upon is vast and diverse in nature. I draw from the literature in mathematics education broadly, as well as STEM education, learning sciences, psychology, and organizational theory. I also draw extensively on my own research. Early in my career I focused on the value of peer critique and novel forms of assessment for learning [219, 220]. For nearly the past decade, I've been co-developing the equity analytics methodology and the EQUIP observation tool [230]. Equity analytics is both a research methodology and approach to professional development. In co-developing this methodology, I have had the opportunity to code and analyze hundreds of mathematics classrooms across K–12 and postsecondary settings. I've also worked extensively with instructors in other STEM and non-STEM disciplines. I've work with numerous cohorts of faculty participants, and I've supported faculty members to coach their peers. I have witnessed firsthand the ways that classroom hierarchies are persistent, pervasive, and common across a variety of settings in the US. I've also seen how the use of data, community support, and targeted teaching strategies can disrupt those hierarchies. I describe how to use these ideas to catalyze your learning in Chapter 7. This book is grounded in a wealth of empirical data and practical experience. Nonetheless, I take a conversational tone throughout, for readability. I hope this book can be useful to anyone teaching mathematics, and expertise in educational research or teaching are not prerequisites to learning from this book. If you are interested in anything in this book or have further questions, I would love to hear from you (daniel@danielreinholz.com).

1.2 Acknowledgements

This book would not be possible without the wisdom, experience, support, and love of dear colleagues, friends, and collaborators. I have learned so much over the years from others, and I am grateful to them for walking on this path with me. Teaching to disrupt hierarchies is as much as learning process as it is any specific set of practices, and I have found immense value in participating in this process with others.

To my partner Suparna Kure, you have been a source of hope, inspiration, love, joy, and guidance throughout this process. You have taught me so much about what it means to be in community with others as we work to manifest a more hopeful and just world. You are my first, favorite, and forever co-conspirator in transforming education. And with much love to our two youngest co-conspirators, Manzil and Rahi. With love and gratitude, hamesha. Ek din aa gaya hai.

To Niral Shah, it's hard to describe how special our collaboration and friendship is. I feel lucky to know you and to have spent the time we have working together. Your wisdom and ideas permeate this work. Building EQUIP and equity analytics is a collaboration in the truest sense and one that I know is almost impossible to replicate. Anytime I need someone to tell it to me straight, you're my guy. I have so much appreciation for your directness and knowing that you always have my back and will always push me to be better. I hope that our friendship and our work together will carry on for many more years.

To Robin Wilson, I remember meeting many years ago at the RUME conference and being so impressed with your amazing accomplishments as a teacher and activist and how you still carried yourself with so much humility. In the early days of the EQUIP work, I had no idea if these ideas could be taken up productively by math faculty, but you showed me it was possible. You have truly been instrumental to the development of this approach and my own learning. You're one of my favorite people and I love working with you because you keep everything fun and also you bring incredible wisdom to the process.

To Cathery Yeh, am I glad for the serendipity of being in the same STaR cohort! From those early days of getting to know one another, I was amazed by the hope, love, and humanity you brought to the work and to the world. You know how to make everyone in the room around you feel good while at the same time pushing them to be more critical in their thinking. Not many people can do that! I love our collaborations on EQUIP (and beyond) and the ways that we can push each other and be in community together.

To Chris Rasmussen, you are one of the best mentors and friends I could ask for. My road to academia wasn't a linear one, and when I finally made it SDSU, I still had a lot to learn. You've been a support every step of the way in pushing my thinking, sharing new ideas with me, and creating opportunities for me to be successful. The way you pay it forward (both to me and others) shows your humility, kindness, and care for others. I'm grateful to be in your circle as a friend and colleague.

To Rachel Lotan, although we have only met recently, your work has long been an inspiration, as should be clear in this book. Over the past year you have become such an incredible supporter of this work and of me personally, and I am so grateful to you.

I would also like to thank all of the EQUIP coaches and participants who I have worked with over the years. I won't name anybody here to protect anonymity, but y'all know who you are. You are amazing and inspire me and have taught me so much. Thank you for working together with me on this. Many of the insights in this book have come from you.

I'm also grateful for all of my students who have been a part of this work, especially Amelia Stone-Johnstone, who worked with me from the early days of the EQUIP. I also thank Brooke Ernest, Sam Ridgway, Naneh Apkarian, James Giles, Lorenzo Sianez, and Sarah Kouzi. This entire EQUIP enterprise truly wouldn't have been possible without you, and I hope you see the huge impact of all of your efforts.

I would like to give a huge shout out to all of my friends and colleagues who took the time to read a draft of this entire book. I only asked you to do this because I look up to you sincerely and love all of the wisdom that you have to share with our field. Sara Rezvi, Stan Yoshinobu, Juuso Nieminen, Brian Katz (BK), Charles Wilkes II, Aileen Reid, and Deb Carney. I have so much gratitude for all of you. Sara provided targeted feedback on the disability sections, Juuso provided deep guidance on assessment, and Charles offered his wisdom around assigning competence and feedback. I also thank Sam Ridgway for her insights into the disability work, as well as Spencer Bagley and Lynn Cevallos who also shared their wisdom to help me improve the alternate assessment sections. Annie Wilhelm provided feedback on the technical details of statistical methods. I thank Johanna Rämö for sharing practical examples from her expert teaching to be included in the book. I thank Ernesto Calleros and Bill Zahner for their input on language demands of tasks. Thanks to all of you. I have learned much from you and am very grateful for your friendship and collaborations.

I also want to thank Stephen Kennedy, Liz McMahon, and the anonymous reviewers of the *MAA Notes* Editorial Board. Your support, critical feedback, and guidance has gone a long way in shaping this work to make it more rigorous, useful, and supportive of the mathematics community. Thank you so much!

Lastly, I thank the National Science Foundation for providing monetary support for much of the EQUIP work. The work and opinions expressed here are my own, and do not represent the National Science Foundation in any way.

1.3 About This Book

This book aims to be a practical manual with concrete teaching strategies that you can use in your classroom to promote more equitable interactions. Recognizing that classroom inequities are grounded in historical and ongoing systems of oppression in our broader society (e.g., White supremacy, patriarchy, ableism), I want to be realistic about what we can hope to achieve as educators. Even if you were to take every strategy in this book and use it perfectly, it wouldn't suddenly *fix* inequity in education. It would be a step in the right direction, but without addressing the deeply rooted inequities in our society, our classrooms will never be truly equitable. And as soon as your students leave the classroom, they will have to face the oppressive world around them. In this sense, oppression in society puts an upper bound on what we can achieve through equitable teaching. Change in the classroom is absolutely necessary and serves as an important form of harm reduction for our students, but equitable teaching is inherently limited and should not be viewed as a panacea.

Writing this book was a massive undertaking. To the best of my ability, I have attempted to synthesize a large, diverse, and sometimes contradictory body of literature about equitable teaching in mathematics. The way I have approached this is grounded in my own experiences, identities, and research work, which provides a somewhat limited but hopefully useful overview and synthesis of the field. Some limitations of the book result from my own biased, singular perspective, while other limitations are simply the limitations of our field. Overall, mathematics education research has done a much better job of recognizing what is not working, while developing concrete strategies to address inequities is still a challenge. Furthermore, our field often lacks nuance in how we talk about equitable teaching when it comes to concrete approaches. In our EQUIP work, Niral and I have focused on the creation and use of data with social marker specificity (that is, data that can speak to the experiences of social marker groups such as women, students of color, or Black women). Thus, rather than assuming *equitable teaching* is one-size-fits-all for minoritized students, instead, we try to draw attention to the experiences of specific minoritized groups (through data) so that instructors can reflect on them and take up strategies to address inequities. While there are certainly many areas of overlap in different forms of equitable teaching, it would be a mistake to treat equitable teaching as a monolith that will work effectively for all minoritized learners of different backgrounds. While I wish that I could offer concrete guidance at a finer granularity (and sometimes I do), often it is simply not possible given where the field of mathematics education is. Throughout the book, I will use the term equitable teaching to signal that there is a need to attend to and center minoritized students as a part of the teaching process (which is made explicit through data generated by the EQUIP tool). This should not be taken as a sign that equitable teaching is the same for all students.

The primary audience for this book is college mathematics instructors. This includes tenure-track faculty members, research and teaching professors, full-time instructors, adjunct lecturers, graduate teaching assistants, undergraduate learning assistants, and anyone else supporting the teaching and learning enterprise. It is for people teaching at a university, liberal arts college, community college, or elsewhere. I assume that many of the readers may have had little formal training in education, so I try to start from the ground up. Broadly speaking, this book as aimed at practice (not research) although it does provide some useful summaries of research. As such, I aim to discuss the research literature in a way that will be consequential for teaching practice, but certainly not exhaustive from a research perspective.

Although I write specifically for college mathematics instructors, many of the techniques are general and can be used effectively in K–12 schools or in disciplines outside of mathematics. Whoever you are, if you're interested in disrupting hierarchies in your teaching—or in an informal educational space—hopefully you can find something you can use in this book. I focus primarily on face-to-face classrooms, but many of these ideas translate well to synchronous online classes. In various parts of the book I specifically call out strategies for virtual learning environments. These strategies are best suited to a smaller classroom size (less than 100 students), but some ideas will work for larger classes too.

No matter where you are in your teaching practice, I believe that there is something that this book can offer you. If you're a new teacher, or if you've spent your entire career using pure lecture, you probably jump straight into course content on the first day of class, and most of your class sessions involve you solving problems or proving theorems in front of the class. The main form of student interaction is probably students asking questions so that you can clarify your lecture. If this sounds like you, a first step might be for you to use this book to explicitly build community early in the semester. You can also delve into the world of questioning strategies, incorporate opportunities for students to work on parts of problems, share their work with the class, and have interactions with peers (e.g., through a think-pair-share). This type of transformation would move your classroom away from a "pure lecture" to more of an "interactive lecture" or presentation.

Suppose that you already have students involved in your class sessions. You typically work out some example problems and you give your students time to work out problems in small groups and share out their ideas to peers. This book will help you think through how you support your students to engage with meaningful tasks and succeed with cooperative learning. It will also offer strategies you can use to help elicit deeper thinking from your students. It includes strategies you can use to ensure that everyone has a chance to share, not just a few dominant students. As you implement these strategies, you can start moving from an active classroom to an equitable classroom.

Now suppose that you've already been doing this work for quite some time. You've been developing your equitable teaching practices for years, and you're a seasoned educator. Is there still something to offer in this book? Absolutely! This book will provide you with tools—both theoretical and practical—to develop more specific teaching practices to disrupt the hierarchies in your classroom. A key—and unique—value-add of my work is social marker specificity of practices. In other words, beyond simply inviting equitable engagement broadly, you'll develop techniques that you can use to advance anti-racism, gender justice, disability justice, and so forth. These practices can help you consider the participation of specific groups of students in your class. This can help you further refine and enhance your already effective pedagogy.

These examples hopefully give you a sense of the landscape of teaching that moves from instructor-centered to student-centered and equitable. There's no one way to use this book, and there's no single ending point for someone who reads this book. The patterns of hierarchy, and teaching strategies that reinforce them, are predictable; I have seen them play out across hundreds of classrooms. When observing classroom participation, there are measurable patterns based on gender, race, disability, and other axes of privilege/marginalization. The more dimensions along which a particular student is privileged, the more likely they are to participate in—or dominate—classroom discussions. Teaching strategies that allow for this domination are also predictable. When instructors allow students to shout out answers or simply call on the first student to raise their hand without managing participation in a more intentional way, it almost certainly leads to a small number of students from privileged groups (e.g., nondisabled white men) dominating the discussion. This pattern is predictable, yet surprising to most instructors, when they actually see data describing the interactions in their own classroom. With a little bit of time, practice, and intentionality, any instructor can learn to disrupt these common patterns and reduce hierarchies in their classroom.

The book is organized as follows. Chapter 2 provides a brief introduction to learning and mathematics education research. Chapter 3 focuses on inequities and hierarchies in mathematics classrooms. Chapter 4 focuses on creating a productive foundation for learning in your classroom. Chapter 5 is organized around facilitation practices that you can use to influence who gets to participate and how. Chapter 6 concerns how to use assessment to help foster interdependence and independence for your students. Chapter 7 focuses on your processes as a learner, to help you get the most out of this book.

At the beginning of each chapter I offer a brief introduction and guiding question for your reading. At the end of each chapter, I close with a few questions, reflections, or activities for you to think about.

2

Teaching and Learning

This chapter will provide you with an overview of the teaching and learning process and research on mathematics education. Before you read this chapter, I'd like you to take a minute to write down some ideas you have about teaching and learning mathematics, and where those ideas came from. Please take out a piece of paper (or use an electronic device) and take five minutes to write these things down. This extra processing time will make it easier for you to see what is new to you. In addition, this creates an opportunity for you to write down *new ideas* in this chapter as you encounter them. As you read the chapter, consider the reflection question: *in what ways is research about teaching mathematics consistent with what I already know, and what are areas for productive growth and learning to improve as an instructor?* Think specifically about how what has worked well for you as a learner may not always work for all learners.

2.1 Learning to Teach

Despite societal narratives that some people are naturally gifted teachers while others are not, teaching is a decomposable set of practices that can be learned by anyone. If you would like to learn those practices, you have come to the right place. This book is informed by over a decade of my research to create equitable mathematics learning environments.

2.1.1 A Concrete Approach

A few years ago, I was invited to facilitate a professional development course for STEM faculty that had been developed at a prestigious university. The course covered a wide array of fundamental topics in learning theory, and by the end of the year, participants could adeptly discuss foundational principles of learning and connect them to what was needed to build effective learning environments. However, when I finally observed those same faculty members teaching, their teaching practices hadn't changed in the slightest. While the faculty spent a lot of time *talking* about how to change their teaching, there was much less emphasis on *doing* the changing.

This book focuses on concrete teaching practices that you can implement *today*, alongside supporting explanations of the theory and evidence behind those practices. In my experience supporting faculty to learn to teach, starting with concrete classroom practices provides a productive entry point into instructional change. I provide brief summaries of relevant theories throughout the book, but primarily, I will provide you with new techniques that you can use to refine your practice, alongside examples of what it could look like when you implement the techniques. As you transform your teaching through new practices, you will gain a deeper understanding of the underlying theories, which will help refine your practices, and so forth.

This is the approach that I have used with numerous college faculty members to help them transform their teaching. I provide coaching in both one-on-one and community-based settings. I provide regular feedback through concrete data on classroom participation patterns. In between observations, I help instructors adopt new practices, providing them with means to shift the observed participation patterns. My approach to change is incremental. Instructors adopt one new practice at a time, and watch the impact on their data. A year later, the instructors have made impressive changes to their teaching practices—which can be documented with empirical data—and they have developed deeper understandings of underlying theories. Transforming your teaching is most effective (and fun) when you have a community of supportive colleagues or a coach, but even if you pick this book up by yourself, you're sure to find something of use.

2.1.2 Learning through Practice

Although lecture is ubiquitous in universities, the lecture model doesn't reflect how people learn through everyday experiences. Stereotypical images of a university depict students sitting quietly in desks while a professor (i.e., the sage on the stage) spreads knowledge to the students. The reality in classrooms—especially mathematics classrooms—is not far from this stereotype. Why do we teach in a way that is so disconnected from how people learn through everyday experience, and what the research literature tells us about teaching and learning?

Try to think back to the last time you tried to learn a new skill, such as cooking, playing a sport, learning an instrument, playing a boardgame, solving a puzzle, etc. You may have watched others do this activity, and someone else may have explained to you how to do it. All of this helps set the stage for your learning, but at some point, you have to do it for yourself or you won't really learn anything. After you start practicing, you may be able to see some of the areas where you need to improve. Even better, you might receive feedback from someone who has more practice than you at the activity. To be clear, you can't learn something without trying it for yourself!

Watching cooking shows on repeat doesn't mean that we can prepare a gourmet meal. Listening to a concerto does not mean that we can play piano. Watching professional sports does not mean we can make the game-winning pitch. Being an avid reader doesn't mean we're ready to produce the next New York Times bestseller. Watching Simone Biles take the gold medal doesn't mean we're ready to compete in the Olympics. In fact, we could watch (or listen to) these performances for our entire lives, and it would do little to prepare us to reproduce them.

Many of the mathematics classrooms I have observed—either as a student or as a faculty developer—have relied heavily on expert modeling (i.e., lecture or direct instruction), provided some opportunities for independent practice, and provided little meaningful feedback (just grades and a few written comments). While these are all ingredients of learning, the recipe is flawed. Most cake recipes include salt because it helps enhance the flavor of other ingredients. But a cake with too much salt is an inedible nightmare. Lecture is like salt. In small proportions, it is a valuable ingredient, but in excess, it's a disaster.

A more useful model for teaching is that of a coach or personal trainer.[1] Suppose you were taking piano lessons. You would certainly want an instructor who is excellent at piano (i.e., a performer). Equally important, you would want someone who knows how to organize the learning environment to help you practice effectively (i.e., a teacher). And you would want feedback on your practice. Although your teacher would model exercises from time to time, and it would be fun to watch them perform, this would only be a small part of the process. It's for this reason that professional musicians do not necessarily make good music teachers. While they can play very well, they might not organize effective practice for you, and their feedback might not be insightful. Conversely, although a music teacher does need to be relatively competent with music, they may never reach the pinnacles of musicianship. The skills of teaching and performance are related but distinct.

In this book, we will focus on those three main ingredients to teaching: instruction, practice, and feedback. We won't spend much time on instruction because you're probably already a very competent lecturer. We will focus quite a bit on how to create learning environments that promote effective practice. I use the term practice to refer to two related but distinct meanings. Practice often refers to repeatedly engaging in an activity as a means of improvement. Practice also refers to something that one does within their domain of expertise (i.e., a mathematical practice), such as making conjectures, proving theorems, or modeling phenomena. We will discuss how to incorporate authentic practice into your class sessions, to help students develop more effective ways of practicing mathematics as a social activity. Lastly, we'll discuss how to produce meaningful feedback for students.

[1] I recognize the inherent tension between effective teaching and institutional pressures of reducing costs and "higher efficiency." Sometimes this requires us to teach in large lecture halls with hundreds of students, which means we need to adapt effective teaching practices to a less-than-ideal circumstance.

Like your students, you will need effective instruction, practice, and feedback. This book is one source of instruction, outlining theoretical ideas and teaching practices you can implement. In order to solidify your learning, you will need to actually try implementing these practices. Ideally, you would have access to a coach or learning community to provide feedback to you. But if not, you can still gather input from students, faculty peers, or even by video recording your own teaching to generate feedback for yourself.

Sometimes to improve your teaching practice you may incorporate different educational techniques that are associated with "buzzwords." You may have heard terms such as active learning, flipped classrooms, or inquiry-based learning. I've heard instructors say that they have "done a flipped classroom" and based on their experiences they conclude "flipped classrooms do not work." Realistically, that instructor probably lacked the necessary skills and practices to make a flipped classroom work. Learning to use new techniques effectively takes time and practice; it doesn't happen overnight. Active learning, flipped classrooms, inquiry-based learning, and so many other paradigms can be effective, if used well. This book isn't grounded in any specific teaching paradigm and can be used in any setting in which students are active participants in the learning process.

2.2 A Brief Introduction to Mathematics Education Research

As we begin our journey to understanding how to disrupt hierarchies in mathematics classrooms, it is helpful to have a general understanding of the field of educational research, how it has evolved over time, and what types of claims can and cannot be made about teaching and learning. In this book, I draw widely from fields related to mathematics education, educational psychology, sociology, and so forth.

2.2.1 Research Paradigms

A brief history of educational research over the past hundred years helps us understand what the field has learned about learning, and how certain we can be about what we have learned. In the early 1900s, educational research was organized around the behaviorist paradigm. Behaviorism aimed to make research as "scientific" as possible, by eliminating constructs that couldn't be measured, and instead focusing on a stimulus (the thing that happened) and response (the outcome). A canonical example of behaviorism is Pavlov's dogs. As the story goes, before Pavlov would feed the dogs, he would ring a bell. Over time, the dogs learned to associate the bell with food, whether any food was present, and they would begin to salivate as soon as they heard it. In this case, the stimulus was the bell, and the response was dogs salivating, in anticipation of food. Although it is rather simplistic, associating the ringing of the bell with food can be taken as a signifier of learning. Behaviorist research had its roots in studies of animal behavior, and later the same techniques were applied to human learning.

Two hallmarks of the behaviorist paradigm are a) breaking down complex learning tasks into smaller pieces, and b) using rewards and punishments to selectively cultivate certain behaviors. Because behaviorism focuses on stimulus-response, the goal was to decompose learning into a series of stimulus-response associations, which constitute the larger skill. By practicing each small part, the idea was that the learner would master the whole (e.g., repeatedly practicing arithmetic skills to lead to larger conceptual understanding). The use of rewards and punishments relates to extrinsic motivation. Examples of this are all around us in academia: merit based raised, student evaluations of teaching, the whole system of "publish or perish," posting articles on social media and hoping for "likes," or having a large number or views on a YouTube video. In teaching, the most obvious example of behaviorism is grading. When students perform in the way we would like them to, they are rewarded with a better grade. When they do not (e.g., showing up late or missing classes), they might be punished with a lower grade. As educators, we are using a particular stimulus (a reward or punishment) and attempting to elicit a response (better compliance or performance).

Research in behaviorism continued for many decades, but eventually, researchers began to uncover the many ways in which these theories were incomplete. The theories didn't account for things such as motivation or cognition, and as a result, behaviorist theories were not able to explain the results of new research studies. In modern times, behaviorism has fallen out of favor as a research paradigm, and is often looked down upon as outdated, oversimplified, punitive, and reductionist. Ironically, behaviorist ideas pervade the education system and the modern working world. Rather than viewing behaviorism as *wrong*, I think it is most productive to view it as *incomplete*. Rewards and punishments do shape behaviors, but they also have unintended consequences that can potentially undermine their purpose in the first place (e.g., external rewards can undermine intrinsic interest in a subject [67]).

The next big paradigm shift in education research was towards constructivist and cognitive approaches, in the 1960s and 1970s. This shift was largely the result of pioneering work of Jean Piaget and his genetic epistemology [212]. Unlike behaviorism, which focused on measurable behaviors, constructivism focused more on abstract ideas in the head, like knowledge, cognition, motivation, self-efficacy, self-regulation, and so forth. These new ideas allowed research studies to be interpreted in more robust ways. A key tenet of constructivism is that students are not blank slates, but that all new knowledge must be built on prior knowledge. As a result, teaching is not just transmitting information, but requires creating a learning situation that helps draw out the resources that students already have in a way that helps them develop new conceptions. The existing conceptions are not "wrong" ideas, but may be productive ideas being applied beyond the boundaries in which they make sense [273]. An instructor's job requires understanding those existing conceptions so that they may be refined. Educational studies that aim to enhance and measure learning are typically framed in some sort of constructivist paradigm. Many of those studies are less recent, because in the past few decades other paradigms have emerged and are used with increasing popularity.

In parallel to Jean Piaget's work, Lev Vygotsky's work fostered a new paradigm in the Soviet Union. This sociocultural approach didn't catch on in the West until the late 1970s, with a translation of a collection of Vygotsky's works, called *Mind in Society* [306]. Vygotksy was concerned with how learning is situated in social contexts, specifically how artifacts mediate the thinking and learning process. An artifact could be a symbol, sign, language, physical object, or simply anything that can be used to hold a specific meaning. Consider one of the foundations of mathematics: a conception of number. Studies suggest that while infants can innately distinguish between very small numbers like 1, 2, 3, or maybe even 4, conceiving of higher numbers requires mediation [198]). In other words, humans can only intrinsically distinguish between very small quantities, without the help of a mediating tool. These mediators could be fingers used for counting, language (i.e., words for one, two, fifteen, twenty-five, and so forth), or a number-line like representation. While there are many possible mediators, to develop a sense of number that allows us to conceive of and distinguish larger quantities, research strongly suggests that humans need to internalize such representational forms. Without mediation, larger numbers simply wouldn't make sense to us.

As another example, consider modern mathematical notation. In antiquity there was no such notation, and algebraic problems were written in words (a so-called rhetorical algebra). Algebra problems written in words can be quite difficult to solve, and sometimes even to comprehend.[2] In contrast, when such problems are formulated using algebraic notation, they can be solved quickly and efficiently. It is not just a matter of representation, but algebraic notation fundamentally changes how we think of structures and relationships. The importance of representational forms and how they influence our cognition is a cornerstone of Vygotsky's work and sociocultural perspectives more broadly. The social view of learning has broad utility and has allowed researchers to understand how learning is socially situated in much greater depth.

The social turn of education research became widespread in mathematics education in the 1990s [123]. A foundational contribution to this work was Lave and Wenger's description of learning as participation in social practices [144]. From this perspective, learning is not understood as an accumulation of knowledge in the mind. Rather, learning is expressed by one's ability to participate in social activities that are specific cultural practices. In the case of mathematics, this would constitute solving problems, making conjectures, constructing proofs, modeling phenomena, and so forth.[3] Initially, learners are only able to participate in these practices in a peripheral way, but over time, they become more deeply ingrained in the community and can become meaningful participants. With the uptake of sociocultural approaches, the field of education has become much less concerned with measuring learning, and instead attends to the myriad ways in which the teaching and learning processes are tied to social processes.

In practice, many researchers draw upon both constructivist and sociocultural paradigms to understand the teaching and learning enterprise [263]. Neither perspective is "correct," but each of them draws attention to different phenom-

[2]Furthermore, when one is confined to representing the mathematics through concrete "rhetorical" situations described in words, the notion that an imaginary number could represent a solution was incomprehensible. In this way, modern notation allowed for new ways of thinking which created opportunities for further mathematical discovery.

[3]As should be clear, what constitutes a mathematical practice is historically and socially situated. With the advent of computers, entirely new fields of computation in mathematics have opened up.

ena and allows for different explanations of learning. Overall, K–12 mathematics education now leans more heavily towards sociocultural paradigms, whereas Research in Undergraduate Mathematics Education (RUME) is still more situated in constructivism (but this is a generalization). In recent decades, sociocultural work has increasingly taken on political dimensions, focused on constructs like power, race, and gender. Given the shift away from trying to measure learning, modern research in education does not necessarily speak to questions of "what works" in the ways that mathematicians might prefer. Instead, research provides strong theoretical tools for understanding the depth and complexity of learning, which rarely leads to simple answers.

2.2.2 The Role of Theory

While research paradigms provide broad ways of thinking about how learning happens, they don't specify how to design optimal learning environments. For example, constructivism attends to students' prior knowledge, but that alone doesn't tell us how to productively build on that prior knowledge. Rather, we would need some type of theory—such as knowledge in pieces [71]—that actually specifies in depth how students reorganize prior knowledge to construct new meanings, how knowledge can be reorganized in response to certain stimuli, and so forth. Similarly, while a sociocultural paradigm helps us attend to the ways that learning is constructed in social contexts, it is too broad to directly inform instructional design. However, theories describing phenomena such as status [54], identity [117], formative assessment [20], or stereotype threat [280] can provide more concrete insights into how we create our learning environments. Still, these theories don't directly translate into specific tangible classroom actions, and that is where some sort of design research comes into play [48]. Such research takes theoretical claims, uses them to design instructional interventions, and the subsequent study of those interventions speaks back to theory [222]. In short, research paradigms give us big picture ways of thinking about teaching and learning, but educational theories (grounded in specific paradigms) are what we use to make hypotheses that can inform design research about specific methods of instruction.

Theory plays an important role in the development of generalizable knowledge. Much like in physics, a properly designed experiment can be used as a mechanism to generate further support for a theory, refine that theory, or even upend an existing theory. As such, a strong theoretical grounding is the basis for good empirical work that teaches us things about teaching and learning. The sociological concept of *status* provides a useful example of such theory building [49]. Status refers to how students are perceived in the classroom, both by their teacher and other students. Classroom status is often conferred through factors such as prior academic achievement, popularity, language proficiency, and social attraction.[4] Early empirical studies—mostly based on laboratory settings—documented how status is linked to participation [16]. Namely, students perceived as having more status are more likely to dominate conversations. A next step was to design an experiment in which status could be modified and observe the resulting effect on participation. Researchers found that modifying status did indeed impact participation [51]. This laboratory work was later taken to classrooms, and multiple status interventions were developed to equalize student participation (see Section 2.3.7 for more detail on status interventions) [52]. Research documented how these interventions subsequently impacted student participation, and also student outcomes [49]. When researchers designed studies focused on status, their work was grounded in sociological theories, and the design studies aimed to translate this theory in educational practice. The researchers created a causal link between status and learning because status mediates participation, and participation promotes learning. This provides a useful example of theory building for instructional change [156].

In contrast, research that is not grounded in theory does not contribute to generalizable knowledge in the same way. Suppose that a researcher does a comparative study of two instructors, observing their classrooms, documenting teaching practices, student outcomes, and so forth. The instructor finds that students in one of the classrooms did better and infers that the differences can be attributed to teaching practices. While this might be true, there are also

[4]When I think back to high school, I remember a variety of cliques (e.g., jocks, preps, punks, skaters, stoners, nerds, geeks), including the students who were seen as popular, and those who rejected from popular circles. I recognize that these are also culturally and historically bound cliques, so you may need to substitute other cliques based on your experiences. The popular students were perceived as "high status" because they were looked up to or envied by other students. This conferred benefits on them such as increased opportunities for social engagement, and increased self-esteem, and reduced consequences for misbehavior (i.e., they could "get away" with more). Similarly, we can think about how some students are perceived as higher status than others in a mathematics classroom. While popularity may be a component of such status, the social perception that one is "good at math" also increases status and means that peers are more likely to listen to and elevate the ideas of those students.

a wide variety of confounds that could be impact student differences in outcomes. Moreover, even if this was true, the researcher would know little about why those practices worked, and it wouldn't help contribute to generalizable knowledge in a meaningful way. In other words, when research isn't guided by theory, it can be too difficult to separate the signal from the noise.

Overall, educational theories do lead to more effective teaching, but it is not as straightforward as "technique X will lead to outcome Y." A more holistic picture requires understanding *what* works and for *whom* and *why* it works and understanding *how* to make that happen in complex sociopolitical contexts. Learning is complicated, classrooms are complex systems, and transforming practice is nontrivial.

2.2.3 Forms of Evidence

As a social science, the forms of evidence used in educational research are vastly different from those used in mathematical arguments. While a correct proof can validate a claim for all times and all places, there are no proofs in educational research. Even when compared to the physical sciences like physics and chemistry, the claims made by educational research are much more tentative. Claims also tend to be bound to a specific time, place, and historical context, which makes it more difficult to generalize across settings.

With these caveats aside, there are still many things that the field of education can be quite certain about. Evidence doesn't come from a single study alone (like a single mathematical proof), but rather, represents the accumulation of knowledge across a wide variety of studies. Each time a study provides more confirming evidence to support existing theories about how teaching and learning work, it provides a greater level of certainty for education researchers. For example, research from a sociocultural perspective has provided great insights into how artifacts mediate thought (and thus teaching and learning). These contribute to a consensus within the field of education that learning is a social process of meaning making. Clearly, any account of the learning process that did not account for these social aspects of learning would be incomplete.

Making causal inferences from social science research is a complex practice that has its own field of methodological study. The statistical methods underlying research can be quite complicated, and the complexity of quantitative methods continues to grow. When interpreting research studies, having some familiarity with statistical hypothesis testing is very helpful, but goes well beyond the scope of this book. One key idea used to support or reject hypotheses is a p-value. The p-value describes how inconsistent a set of data is in relation to a particular statistical model (e.g., the null hypothesis, which states the outcomes are the same for two populations) [183]. A common, albeit somewhat arbitrary, cutoff for p-values is that any situation where $p < 0.05$, results are considered to be statistically significant (i.e., unlikely to be due to chance, but due to an actual effect). Of course, a small p-value doesn't guarantee there is a real effect[5], merely that there is less of a fit between the data and the null hypothesis. Unfortunately, p-values are misunderstood by many,[6] and a number of statisticians would argue for other approaches to do research, such as Bayesian statistics. Despite these complexities, the use of p values is still common and well-accepted in education research, so I will provide them throughout the book when they are available.

Typically, the gold standard for drawing inferences is a randomized controlled experiment. With a sufficiently large sample size, randomization helps account for a wide variety of biases that might be present in the samples and other intervening variables. For example, when testing the efficacy of new drugs, large scale randomized controlled trials (RCTs) are the methodology of choice. Although RCTs are used in education, they are expensive and time consuming. Also, it is often not easy to randomize in education studies (e.g., you can't force a teacher to teach in a particular way, and if you could, they might not implement with fidelity). In practice, education often relies on quasi-experiments with a comparison group, or some sort of matching procedure such as propensity score matching to help ensure that the comparison is a valid one (and that effects aren't simply due to differences in the original samples). Pre-tests are often used to show that comparison groups have similar performance. All these methods can help, but they are still weaker than randomized experiments. Educational researchers also use a wide variety of observational methods which can also be useful but are subject to even more potential biases. To keep a long story short, it isn't always easy to draw a

[5]And smaller p values don't mean stronger effects!

[6]This is such a large issue that the American Statistical Association released a statement on p-values [311]. Interested readers will also find more information by reading [183].

strong inference from a single study given methodological limitations, practical constraints, and the high cost of doing research in a society where education is woefully underfunded.

2.2.3.1 Effect Sizes

Another barrier to drawing conclusions across studies is that most studies use their own unique outcome measures. To help mitigate the issue of comparing studies, we can use effect sizes. An effect size can be thought of as a quantitative measurement of the strength of the relationship between two quantities. It is essentially a weighted mean difference. The most common effect size in education research is Cohen's d, which is computed as the difference in means divided by the pooled standard deviation. The two means could be for different populations, or it could be the same population measured at two different points in time.

$$d = \frac{M_1 - M_2}{S_p}$$

Another common measure is Hedge's g, which is essentially the same as Cohen's d but it uses a pooled *weighted* standard deviation for the populations, which is useful when the samples are significantly different in size. Sometimes, researchers will also include a bias correction for smaller sample sizes ($N < 50$). While there are differences in these details, both Cohen's d and Hedge's g are simply standardized ways to look at differences in two populations or conditions.

These measures can be computed for a wide variety of different outcome measures, allowing the results from different studies to be compared. Of course, one still should have some caution in comparing studies, given that the studies may have different durations, participants, assignment to groups, and so forth. Our field often relies on guidance from Cohen to interpret the meaning of effect sizes (with $d = 0.2$ as small, $d = 0.5$ as medium, and $d = 0.8$ as large), even though Cohen only provided these as rough guidelines and urged interpretation in context.[7] Throughout the book I provide effect sizes when they are available. An effect size is better than nothing, but making comparisons between studies is not as straightforward simply looking at which one has the larger effect size. Comparisons need to account for context.

2.2.3.2 Meta-Analyses

One strategy that researchers use to draw conclusions about a body of research is to aggregate the findings across a wide variety of smaller studies. This quantitative approach is called a meta-analysis. To perform a meta-analysis, authors conduct a systematic review of the literature, develop inclusion and exclusion criteria for studies, create standardized outcome measures for studies (i.e., effect sizes), and aggregate the effects of studies. A meta-analysis is useful because it helps support more generalized claims from research. Still, one should interpret meta-analyses with caution. Unless researchers take care to only include high-quality and rigorous studies, it is very easy to include a bunch of poorly designed studies and proclaim an effect with great confidence.

As an example, researchers might wonder about the effects of feedback on learning. To understand these effects, a researcher would likely want to distinguish between different types of feedback, which would have different effects. Otherwise, they would be adding apples and oranges together. A researcher might also focus on a particular discipline (like mathematics) or grade level (e.g., kindergarten is different from graduate school). As should be evident, there are a wide variety of judgments made on which studies to include and how to combine their results. A rigorous meta-analysis can be very useful for drawing statistical inferences, but if a researcher isn't careful, their conclusions might not be valid at all.

While meta-analyses can be useful for making general claims, collapsing over a variety of studies also loses nuances. While studies may show that active engagement supports learning overall, how that looks in an elementary art class is very different from an advanced mathematics class. For this reason, meta-analyses must also be taken with a grain of salt. Given variations in contexts, rarely can one claim "if you use teaching technique X, then result Y will occur." An alternative approach is to use logics like design-based research [48] or improvement science [31], which attempt to generate insights by using theory to account for variations in context, rather than using statistics to average out such

[7] We can use the same guidelines for interpreting Hedge's g.

differences. These are useful approaches that I have also drawn upon in my own work, but for our purposes here of establishing some baseline claims about teaching and learning, meta-analyses will still be useful.

In Table 2.1 I summarize the effect sizes from meta-analyses that are included in this book. To be clear, effect sizes like this are not directly comparable to one another given differences in the context, timescale, methods, and so forth, between studies. The number of studies and the quality of studies in a meta-analysis can vary. In general, having more studies would lend more certainty to the results of the analysis, but only if they are higher quality studies. For example, one study of active learning in college STEM classrooms [89] has a large sample but many of the studies are quasi-experiments with no randomization. In contrast, the best evidence syntheses of mathematics interventions in K–12 schools [271, 272] have smaller samples but had stricter inclusion criteria. As such, it is not clear-cut that one of these studies would provide stronger evidence than the other. Moreover, the methods used to compute an effect size vary (e.g., fixed- vs. random-effects models). Because there are so many decisions that go into running a meta-analysis, any findings should be taken with a grain of salt. I don't delve into all the details of methodological techniques for meta-analyses because is well beyond the scope of this book and not needed for our purposes.

Despite the above caveats, meta-analyses can give us a general understanding of impacts. For example, the effects of $g = 0.70$ for Practice Testing and $g = 0.55$ for self-explanation highlight that these active studying techniques are much more effective than passively reading notes. Similarly, the effect sizes for Active Learning in College STEM ($g = 0.47$), Collaborative Learning in Elementary School ($d = 0.33$), and Collaborative Learning in Middle/High School ($d = 0.18$) all point in the same direction giving classroom-based evidence for the importance of active student engagement in promoting learning. The fact that each of these effect sizes is different probably says more about the nature of the studies that were synthesized than about the contexts in which student engagement is more or less important (i.e., there's no theoretical reason to assume that active engagement would be least important in middle and high school, as compared to elementary school or college). Now, if in contrast, one of these meta-analyses had an effect in the opposite direction (showing pure lecture was more effective), then we would have to pause and reconsider.

Topic	Num. ES	Statistical Model	Pooled ES	Citation
Practice Testing	272	Random Effects	g = 0.70	[2]
Self-Explanation	69	Random Effects	g = 0.55	[18]
Active Learning in STEM	255	Random Effects	g = 0.47	[89]
Collaborative Learning (Elem.)	36	Median (Best Evidence Synthesis)	d = 0.33	[271]
Collaborative Learning (Mid/High)	22	Fixed Effects	d = 0.18	[272]
Stereotype Threat (Race)	23	Random Effects	d = 0.52	[182]
Stereotype Threat (Gender)	47	Random Effects	g = 0.22	[86]
Formative Assessment	126	Random Effects	d = 0.29	[145]
Feedback	994	Random Effects	d = 0.55	[318]
Grades vs. Feedback	71	Fixed Effects	d = 1.14	[135]

Table 2.1. An overview of the meta-analyses reviewed in this book. Each row shows the number of effect sizes (Num. ES) that were pooled together to one overall effect size.

To be clear, meta-analyses are a powerful methodological tool, and they are one that can be used to accumulate evidence that may support or refute a theory within education. However, when meta-analysis is used in a way that ignores any theoretical conceptualization of how learning happens, it can also become a process of simply crunching numbers in a way that isn't that meaningful. For this reason, we will want to interpret any results from a meta-analysis in light of what we already know generally about learning, rather than simply taking the numbers at face value without a theoretical explanation.

In this book, some claims can be made with more certainty than others. The basic model of how people learn is well supported by a vast array of studies, but some of the finer details may have more or less support. When meta-analyses are available, it usually means that the topic has been widely studied in the field and there is strong consensus in terms of what the research says.[8] In other cases, there may be multiple studies that all provide confirming evidence in favor

[8]Even though randomized controlled trials are the gold standard that we would strive for in medical research, we won't always have RCTs to

a phenomenon, but it hasn't been widely studied to allow for meta-analysis (or the meta-analysis simply hasn't been done). This still provides some useful evidence, especially if the conclusions drawn from the study are consistent with our overall theory of learning and my own experience teaching and professional development. One final challenge is that a lot of useful educational research and evidence comes from qualitative studies, but they cannot be easily accumulated in the same way as quantitative studies.

Throughout the book, I also draw on my own teaching experience and professional development work. In a few cases, I'll refer to strategies my colleagues have used that I have not, and I'll be clear when I'm doing that. While not all this work has been published in academic journals, I still have evidence for the claims I'm making. In short, I'm not trying to sell you anything that I don't have evidence for myself and haven't used in my own teaching and coaching practices. On the whole, I take a rather skeptical approach to the research in the field, focusing on what we can and cannot infer validly from research.

2.2.4 Summary

In conclusion, there is a lot that can be learned from educational research, but we also must be mindful of the nature of research in complex social systems. Not all research findings are equally valid, and they won't necessarily apply across all contexts. For this reason, a more holistic view of the field (rather than cherry-picking a few studies to make a point) is needed to truly understand what is known about teaching and learning in mathematics. While I obviously can't review all that research here, I do my best to provide a balanced view of the state of the art of what is known about teaching mathematics to disrupt hierarchies, specifically in postsecondary contexts.

2.3 To lecture, or not to lecture, that is the question

There is a narrative in the mathematics community that pits traditional lectures against student-centered forms of learning.[9] As the story goes, on one side of the debate are mathematicians who believe that students must develop fluency with basic procedures before moving onto deeper conceptual thinking. They believe that lecture is an effective way to help students achieve this. On the other side, reformists believe in collaborative learning activities, like peer tutoring, collaborative group work, and whole-class discussions. They focus more on the process of learning than any particular outcome. Reformists accuse the traditionalists of caring more about themselves than their students, and that they promote "drill and kill" over meaningful understanding. Traditionalists retort that reformists aren't real mathematicians, they don't really teach, and they expect the students to do all the work for themselves. Hostility between these two sides has existed for decades, and it has become so heated that the debate has been dubbed "the math wars" [258].

The above description of two camps is false binary, and most mathematics instructors don't hold onto their position so fervently that they are "at war" with others. I believe that these arguments are harmful for everyone in our community, especially the students that we teach. Arguments on both extremes are clearly false, and both communities have more common ground than is sometimes portrayed. Research is clear that procedural fluency *does* support conceptual understanding [62]. Although instructors may disagree about how much time to spend on procedures, how to teach them, and which procedures are most important, we should all be able to find common ground that at least some level of procedural fluency is foundational to mathematical understanding. Similarly, research is clear that the process of learning is crucial to learning outcomes [73], and also that there are many ways to structure a learning environment, but pure lecture is among the least effective.

People who study mathematics education and those who study mathematics need not be at odds with each other. Everyone in our community has something to offer to help improve the lives of our students and the usefulness of mathematics for our society. I believe that overwhelmingly, mathematics instructors want to teach effectively for their students, and as a community, we want to help more people learn how to do more mathematics better. If we put the hyperbole and divisive rhetoric aside, we can all do much better as instructors; we have a lot to learn from each other.

refer to for all aspects of teaching and learning. Part of the issue is that certain phenomena would be very difficult to study in such a setting due to challenges in randomizing, as described above.

[9]In this book I take a broad view of the mathematics community, inclusive of pure and applied fields, mathematics education, mathematics teachers, people using mathematics in industry, and so forth.

Perhaps the largest point of contention in our community is the use of lecture. The problem is often framed as whether to lecture, or not to lecture. I believe that a more productive framing is to ask how much to use direct instruction (i.e., lecture), and when. On one end of the spectrum, there is teaching that is based purely on lecture, with no student interaction. On the other end, there is teaching that is entirely student guided, with no support from the instructor. Research suggests that effective practice lies somewhere in the middle.[10] This means that instructors shouldn't *not lecture*, but they also shouldn't *only lecture*. Research is conclusive that student learning can be enhanced by a variety of active engagement techniques, but to be clear, most student-centered classrooms still include direct instruction (or lecture) to scaffold student learning, provide feedback, clarify explanations, and so forth. Even as an instructor who identifies relatively strongly with student-centered instruction, never in my career have I taught a course that includes no lecture. When it comes to answering the question posed in the title of this chapter, an acceptable answer would be "yes, and..." As an instructor you need not give up your lectures, but you do need to do lots of other things too (which means that lecturing for an entire class period is out the door). Here I provide a brief overview of some of the evidence in favor of a "yes, and" approach.

2.3.1 Attention Reduces over Time

There's an often-stated metric in the literature that attention spans are significantly reduced after 10-15 minutes of a lecture [58]. Although this metric is stated with conviction, it isn't backed up by rigorous empirical research [24, 316]. Even though students' inability to focus on a long lecture is an often-cited reason for not lecturing, student focus during lectures isn't that well studied. One barrier is that it is somewhat difficult to study attention in a lecture. A typical study approach is to embed prompts at some time intervals throughout a lecture to ask students if they are focused or if their mind was wondering, and these prompts may be matched with some sort of performance measure. Here I review two studies that used such methods, which suggest that attention (and consequently learning) reduce over time within a lecture [32, 240].

In the first study, student attention and learning from a 60-minute lecture video was studied [240]. A total of three different lecture videos were used (obtained from Open Yale Courses), from either psychology, classics, or economics. Students were assigned randomly to watch one of the three videos, and were given mind wandering probes at 5, 25, 40, and 55 minutes. A limited number of probes were used so that students wouldn't unintentionally hyperfocus because of continuous prompting. For analysis, student responses to the 5- and 25-minute prompts were combined, and the 40 and 55 minutes were combined, for an aggregate *first half* and *second half* of lecture categories. Students responded yes to mind wandering prompts significantly less often ($p < 0.05$) during the first half of the lecture (35%) than the second half (52%). Additionally, when students were asked questions based on the lecture material, they answered significantly more correctly from the first half (71%) than the second half (57%). The authors replicated these findings in a second study using the same materials [240], adjusting the prompts to slightly different times, (2- and 20-minutes) and (35- and 50-minutes). The authors found similar results in a third, similarly designed study [84].

In another study, students in three chemistry courses were provided with clickers and were asked to report instances of mind wandering and how long it lasted [32]. In this study, a variety of active teaching methods were used in addition to lecture, so the average lecture length was only 12 minutes. As expected, students pressed the mind wandering button throughout the lecture segments. There were two key findings. First, students pressed the mind wandering button significantly more during lecture than during activities. Second, students were more focused during lectures that immediately followed periods of active engagement. These results suggest that by breaking up the lecture with active engagements, students had increased attention span for future segments of lecture.

In conclusion, these studies suggest that attention reduces over time (as one would intuitively expect), and this can reduce student performance. In addition, they provide some preliminary evidence that breaking up lectures with active engagement can improve focus for subsequent segments of lecture. Nevertheless, one would expect that there is heterogeneity in attention based on the context, students, interest in material, and so forth. Thus, the literature is not conclusive about a particular length of time after which student attention wanes, but does suggest that it can happen after a relatively short segment of lecture (even in segments of only 12 minutes). In addition to considering how

[10] To be clear, there is no research that can tell us exactly where in the middle would be optimal, and in practice, that will differ from instructor to instructor and lesson to lesson.

attention reduces over time during a lecture, it is important to consider the potential learning benefits due to active engagement during class sessions. Fortunately, this has been very well studied, as I present in Sections 2.3.2, 2.3.3, and 2.3.4.

2.3.2 Findings from Laboratory Studies

The field of educational psychology focuses primarily on studying learning in controlled laboratory environments. Random assignment of participants to an experimental condition (the new learning technique) and comparison condition (typically extra time studying or another learning activity) helps eliminate any confounding factors, so a particular learning strategy can be studied in isolation. Psychologists have conducted tens of thousands of experiments of this nature, and as a result, we can be quite confident about certain aspects of the learning process. Of course, real classrooms are more complex than a laboratory, so we must consider other variables when implementing these ideas in practice. Nevertheless, laboratory studies provide extensive support for some learning methods over others. Moreover, when these same phenomena are studied in classrooms, which is messier and more expensive, the same general results tend to hold true, providing confirming evidence. Here I provide a brief overview of some of this research.

Dunlosky and colleagues [73] reviewed the evidence in favor of ten common learning techniques, by drawing on meta-analyses of thousands of laboratory studies. The ten techniques they reviewed were: 1) elaborative interrogation, 2) self-explanation, 3) summarization, 4) highlighting, 5) keyword mnemonics, 6) imagery use for text learning, 7) re-reading, 8) practice testing, 9) distributed practice, and 10) interleaved practice. (I define these terms in the paragraphs that follow.) While results from general education settings may not always directly translate to mathematics learning, studies in mathematics that have focused on these same effects have found similar results [186, 241]. The review of ten practices concluded that practice testing and distributed practice were highly effective; interleaved practice, elaborative interrogation, and self-explanation were moderately effective; and the other techniques had low efficacy.

Practice testing involves working on material in a low-stakes way that is not for a grade. This is one of the most widely studied learning strategies in educational psychology, and it is one with extensive evidence documenting its efficacy across contexts. For example, a recent meta-analysis found a medium to large effect size for practice testing ($g = 0.70, p < 0.001$, pooled from 272 effect sizes) [2].[11] These results show fairly conclusively that practice testing is far more valuable for learning than alternative activities like re-reading a textbook or studying lecture notes. What is notable is that in these studies, the solutions to the practice problems need not be correct. In other words, *imperfect* practice is still more beneficial for learning than studying a *perfectly* correct text. In fact, some research suggests that the very act of incorrectly attempting a problem could have learning benefits in its own right [322]. In short, trying to do something—rather than reading or listening about it—is one of the most effective ways to learn how to do it. This is consistent with intuitive ideas about how people learn in everyday experience. [12]

Distributed practice, which is also highly effective, consists of spreading out learning events over time, rather than consolidating them all at once. A canonical example of this is preparing for an exam. Some students may try to do all their studying (or solving practice problems) the night before an exam. As conventional wisdom tells us, this isn't a very effective way to study. Students are much better served by spreading out their studying over time. Distributed practice also increases retention of skills and knowledge over time, not just in the short term [186, 317]. These same effects have also been replicated specifically in mathematics [63, 239, 245]. Because these studies vary in their details and the spacing of distributed practice, we can't precisely determine the ideal spacing that will result in greatest learning or retention. Nonetheless, these studies are another source of evidence showing how when students engage in practice it can significantly impact their learning.

Similarly, *interleaved practice*—when students work on different types of problems at the same time, rather than only focusing on one skill or problem type—also improves performance [243, 244]. Typically, classroom learning is consolidated into chunked events (e.g., students do all homework problems from a single section at the same time). However, a more effective way to practice is to spread the learning events out over time (e.g., students continue to

[11]This recent meta-analysis is just one of the many meta-analyses conducted showing evidence in favor of practice testing. Moreover, as far as educational research goes, high effects sizes are relatively uncommon, which highlights just how striking these effects are on learning, and how they have been replicated repeatedly under rigorously controlled experiments.

[12]Of course, practice testing will be most effective when it is integrated into a larger program of learning that involves teacher guidance, feedback, peer support, and so forth.

revisit problems from a earlier sections in future homework). This type of interleaved practice is relatively easy for us to implement as instructors, and it can help our students learn more.

The other two moderately effective practices are *self-explanation* and *elaborative interrogation*, which both focus on getting students to explain their ideas. Self-explanation studies use a wide variety of different prompts to elicit explanations, whereas elaborative interrogations focus more specifically on asking why-questions to participants. Because these strategies are quite similar, for brevity, I focus on self-explanation effects [42]. The process of self-explanation is simple. When reading an explanatory text, students are prompted to explain the ideas to themselves in their own words. A recent meta-analysis found a medium effect size for self-explanation ($g = 0.55$, $p < 0.001$, pooled from 69 effect sizes) [18].[13]

Like practice testing, students learn more from self-explanation, even if their explanations are not fully correct [42]. When students explain their ideas, it forces them to consolidate their learning in a way that it can be communicated to others, while attempting to resolve potential contradictions in a variety of conceptions they might hold. This act of consolidating and communicating produces more robust knowledge. Similar explanation effects have been found in other settings, such as tutoring [246].

The final five strategies described—summarizing, highlighting, keyword mnemonics, imagery use for text learning, and re-reading—all had low efficacy. These strategies all aim to improve the retention of ideas, but don't include any active processing like the five strategies already reviewed. Let's consider re-reading as an example. As one would expect, research shows that if students read a text multiple times, they typically have better comprehension. However, when one compares re-reading to the more effective learning strategies like practice testing, it simply isn't a great strategy to use. Given that all students have limited time to spend learning, it is helpful to optimize the use of that time with the most effective learning strategies. Overall, these studies provide robust evidence that active processes that involve constructing new knowledge make a significant contribution to learning.

2.3.3 Effective Programs in Mathematics

Now I turn attention to classroom-based studies, which provide a more authentic (but less controlled) context for learning. Most classroom-based studies are not true randomized studies, given the cost and difficulty of implementing such studies. Here, I review two meta-analyses.[14]

The first meta-analysis focused on the impact of 87 studies in elementary mathematics that had randomized or matched samples, were at least 12 weeks long, and had outcome measures that were independent of the intervention [271]. The types of interventions in the meta-analysis broadly fit into three categories: mathematics curricula (i.e., introducing new textbooks); computer-assisted instruction (which provides individualized exercises and feedback for students); and instructional processes (focused on changing classroom instruction). The least impactful interventions were mathematics curricula ($d = 0.10$), and computer-assisted instruction ($d = 0.19$). By comparison, instructional process programs had a larger impact ($d = 0.33$). Although there was heterogeneity in these interventions, overall, they focused on using research-based teaching and learning strategies that helped students engage more effectively with peers in the classroom.[15]

A follow-up study focused on effective middle and high school math programs [272]. This study analyzed 100 studies, in the same categories of: mathematics curricula ($d = 0.03$); computer aided instruction ($d = 0.10$); and instructional processes ($d = 0.18$). Within the category of instructional processes, the most effective interventions focused on students learning collaboratively with peers in specific ways.[16] These results are broadly consistent with findings at the elementary level. In short, the most effective programs were the ones that helped instructors improve

[13]Another meta-analysis of studies only in mathematics corroborates these findings [241].

[14]These studies were "best-evidence syntheses" which is a specific type of meta-analysis that carefully considers context and details of the studies, to provide better interpretation in addition to simply pooling quantitative effects.

[15]Programs in this category implemented a variety of instructional strategies (e.g., cooperative learning, mastery learning, classroom management and motivation, and direct instruction). Of the 36 studies in this category, only 4 of them focused on direct instruction. Moreover, the definition of direct instruction was not "pure lecture," but rather that lessons begin with instructor guidance, and then later focused on individual and/or group practice by students. The programs reviewed, Connecting Math Concepts [313] and User-Friendly Direct Instruction [101] both used research-based strategies (e.g., distributed practice) to enhance learning compared to "traditional direct instruction."

[16]Two types of cooperative learning programs stood out: Student Teams-Achievement Divisions (STAD) and Introduce, Metacognition, Practice, Review, Obtain Mastery, Verification, and Enrichment (IMPROVE) ($d = 0.46$).

their instructional practices to engage students more effectively in more meaningful learning activities (not just creating better lectures). We turn our attention to similar programs in postsecondary mathematics.

2.3.4 Active Learning in Postsecondary Mathematics

In the postsecondary context, student-centered teaching techniques are often described as *active learning*. This means that students are doing some sort of activities or discussions rather than listening to pure lecture. The largest study of active learning is a recent meta-analysis of 225 studies that compared active learning to lecture in undergraduate STEM courses ($g = 0.47$). The findings were striking. Failure rates under "pure lecture" were 55% higher than those observed active learning environments [89]. Largely informed by this study, the CBMS concluded that "the status quo is unacceptable" [255, p. 12]. One noteworthy difference in the post-secondary context is that pure lecture is much more prevalent, whereas most K–12 classrooms (even those classified as traditional) typically include some forms of student practice (even if they are largely individual and procedural).

Early studies of inquiry-based learning (a specific type of active learning that is heavily student-guided) in mathematics provided some limited evidence that active learning could also help reduce inequities [143, 142, 136]. These studies included data from over 100 course sections on four campuses that had Inquiry-Based Learning (IBL) centers. Data included 300 hours of classroom observation, 1100 surveys, 220 tests, 3200 student transcripts, and 100 interviews with students, faculty, and graduate teaching assistants collected at the four institutions over two years. To understand the impact of IBL courses on student learning, the research team sampled three courses with robust samples, and compared IBL students and non-IBL students in future courses. They found that IBL student grades were equal or significantly better than non-IBL students, with modest improvement. Additionally, differences were observed in students' self-reports of learning gains, cognitive gains (e.g., conceptual understanding), and affective gains (e.g., confidence, persistence). The study suggests that students in IBL and non-IBL courses did equally well on traditional tests, but there were some other gains for IBL students, including affective gains, problem solving skills, and increased success in future courses. Notably, women in IBL courses reported greater levels of confidence and intent to pursue further mathematics when compared to peers in non-IBL courses [143]. Despite some promise based on these studies, in Section 2.3.5, I consider a larger body of more recent counter evidence which indicates that unless there is explicit attention to equity, active learning can sometimes produce inequitable results.

2.3.5 Can Active Learning be Inequitable?

Considering the context of failed educational reforms in K–12 schools, it would be surprising if active learning alone produced more equitable outcomes in postsecondary education. For over 40 years, the US education system has focused on reducing racial inequities in outcomes on standardized exams. Two major sets of initiatives have related to the NCTM Standards starting in the late 1980s [187] and the Common Core State Standards of the late 2000s [61]. These reforms largely focused on improving curriculum and teaching methods to help students develop more robust problem solving, conceptual understanding, and explanation skills. These reforms took a *for all* approach, with an ideology that a rising tide lifts all boats. Yet, research shows that these approaches have failed minoritized communities, especially Black learners [162, 17]. In other words, despite the promising mathematical ideas included in these efforts, they had little impact on the inequitable status quo. The evidence gathered to understand these movements is vast in comparison to a few quasi-experimental studies in higher education. To be clear, a *for all* approach doesn't seem to improve racial equity. I argue, as others have before me [162], that equity will only be improved with intentional practices to disrupt inequities. Here I would like to review a few studies that underscore how active approaches can (re)produce inequity.

The first study focused on a research-based model for professional development in elementary schools that was implemented in an urban district for three years [173]. The approach emphasized an anti-deficit view of students, aimed to create status free classrooms,[17] and to engage students in rich mathematical practices with challenging tasks. Teachers in this study were also taught to notice and build upon student thinking. While the approach purportedly attended to equity, it did not attend to race; over three years of professional development meetings, a specific focus on Black or Latinx students never came up. Overall, student outcomes for participating students improved an additional 3.96 percentiles over three years ($d = 0.15$), so the professional development did have a small impact in the aggregate.

[17]This relates to the set of techniques called Complex Instruction, which I review in Section 2.3.7.

However, the impacts were differential by race. Outcome gaps for Black students were exacerbated because of the intervention. In particular, the authors found that Black student performance decreased 4.14 percentiles over three years ($d = 0.15$). This study highlights how something that is better for all in the aggregate, can still exacerbate inequity for some groups of students.

The next study was situated in upper-division undergraduate mathematics [228]. This study was based on the TIMES project, which created inquiry-oriented curricula in differential equations, abstract algebra, and linear algebra. From a larger sample of 42 inquiry instructors and a matched comparison sample, 20 inquiry instructors were sampled because they had assessment outcome data linked to student gender. No statistically significant difference in performance between women in inquiry and non-inquiry classes was present. However, for men in inquiry courses there was significant improvement on the conceptual exams for both differential equations (55.1 vs. $47.7, p = .002$) and abstract algebra (50.4 vs. $42.9, p = .007$), when compared to non-inquiry courses. Moreover, this difference corresponded to a statistically significant difference between men (mean 50.4) and women (mean 35.7) in inquiry-oriented abstract algebra. Thus, while students did better or the same in inquiry courses compared to non-inquiry courses (consistent with Laursen's studies [143, 142, 136]), the impact was differential and favored men, which reproduced inequity. We'll return to this study in more depth in Section 3.3.

A larger field study of elementary schools in four South American countries has findings concordant with these results [9]. Analyzing 10 randomized experiments including over 17,000 students, the program increased mathematics scores (0.18 standard deviations) and science scores (0.14 standard deviations) after one year, and by 0.39 and 0.23 standard deviations, respectively, after four years. However, there were significant gender inequities, with a performance gap between boys and girls growing over the four years, adding up to 0.17 standard deviations of difference in math, and 0.15 standard deviations in science.

The final study was a randomized trial in a military academy [261]. This study involved 29 faculty, 80 class sections, and 1328 students in mathematics and economics. Then, 40 class sections were randomly assigned to flipped or traditional, and each instructor taught at least one flipped and one lecture classroom. In this study, students in the mathematics sections had improved scores overall on quizzes early in the semester (0.3 standard deviations). However, these effects were driven by men, White students, and students with higher ACT test scores. Men in math classes had test scores improved by 0.32 standard deviations, White students improved by 0.385 standard deviations, while racially minoritized students and women received no measurable benefit. On the final exam, there was no measurable effect in the aggregate. However, the flipped classrooms did exacerbate inequities between subgroups. For example, the achievement gap between White students and racially minoritized students (in this case Black and Latinx) was 0.263 standard deviations in the control group, and 0.444 in the flipped classrooms (at 69% increase). Similarly, the difference in quiz means between students with low and high ACT scores was 23% higher in the flipped classrooms. Data on gender were not reported for the final exam.

Overall, these four studies were methodologically robust. Three of four used randomized trials [261, 9, 173], while the other had a modest sample of 20 inquiry instructors and matched comparisons [228]. These studies provide fairly robust evidence showing the ways that active learning can reproduce inequities. Although the research in this area isn't definitive, it is consistent with prior studies showing that active learning improves student learning in the aggregate, but also, it can result in greater inequity if there isn't appropriate attention to equity with social marker specificity (i.e., active learning strategies need to attend to the ways that different groups of students experience the learning environment differently, by social markers such as race). This book provides the tools to make active learning more equitable.

A skeptic could take this evidence and conclude active learning isn't worth implementing given it can cause inequity. However, it is important to note than in most of these studies, the minoritized students did no worse (and some did even better) than their counterparts in comparison classes. In other words, active learning generally produces greater learning for students, but the learning gains may be inequitable (with non-minoritized students learning even more than their minoritized counterparts), unless instructors disrupt the hierarchies inherent in social interactions.

In Section 3.2, I describe concrete mechanisms that create hierarchies in the classroom. Of particular interest will be student participation. Educational theory and empirical data [228] suggests that the inequities that arise in active learning classes can at least be partially attributed to the quality and quantity of participation opportunities afforded to different groups of students. In short, non-minoritized students take up more opportunities to participate, and thus learn, which mediates the production of inequity.

2.3.6 Summary of Active Learning Benefits

I have reviewed a large body of evidence strongly showing why active learning classrooms support better learning. Amidst all these details I want to leave you with three key reasons why participation (or practice) is foundational to learning: 1) practice supports learning, 2) feedback supports learning, and 3) participation promotes identity and belonging.

Practice Supports Learning. As described above, techniques like practice testing and self-explanation are effective because constructing knowledge is more beneficial than memorization. These robust laboratory studies have been further validated in classroom settings [228, 122, 308]. These studies show that classroom participation (e.g., through verbal discussion) has a strong correlation to student outcomes. What these studies make abundantly clear is that students benefit greatly from engaging with peers in social settings to learn mathematics. Such settings provide ample opportunities for students to explain their ideas and learn from the ideas of their peers.

Feedback supports Learning. Feedback is another cornerstone of learning. An large body of evidence documents the important role that feedback plays in enhancing student learning [19, 270, 318]. The value of feedback has opened an entire field of education focused on formative assessment practices, the topic of Section 6.1. Feedback is a crucial part of learning, because it helps learners establish where they are relative to where they are going, and it helps them figure out how to get there [19]. In this way, feedback is the guide map or GPS that allows students to reach the destination of deeper mathematical understanding. Feedback comes from instructors, as well as peers [220]. Peer feedback has been studied across a wide array of studies and is shown to be generally positive for learning [302]. In an active learning classroom, there are abundant opportunities for students to receive feedback from their peers. This happens in pair-share opportunities, small groups, and in whole-class discussions.

Participation promotes Belonging and Identity Development. Over the past few decades, educators have amassed an array of evidence showing how students' sense of belonging is crucial to their persistence and success [98, 109, 165]. Students who don't feel a sense of belonging are more likely to drop out of their program of study [148]. Belonging is enhanced when students form a community of peers who provide mutual support and engagement. When students feel like they belong in an environment, it also helps them develop identities as mathematicians [184]. These practice-linked identities are essential to how learners understand their own positioning with respect to the discipline [185]. In short, students need meaningful opportunities to engage with meaningful content to develop mathematics identities. Simultaneously, forces such as racism and anti-Blackness shape student experiences in ways that exclude many minoritized students [162].

2.3.7 Making Active Learning More Equitable

Although inequities may arise in active learning settings unless instructors explicitly address them, there is a body of research that provides practical tools for doing so. The most widely researched set of techniques with demonstrable effects on student learning is called Complex Instruction [49]. Complex Instruction is a pedagogical approach based on sociological theories and the malleability of status. Students perceived as high status tend to have more opportunities to participate, and their ideas are seen as more valuable. Consequently, targeting status as a site of intervention can be a useful strategy for remedying racial (and other) inequities in math classrooms.

Complex Instruction uses four primary strategies to reduce status imbalances and promote student belonging (i.e., status interventions):

1. The *multiple-ability treatment* involves framing tasks as requiring diverse skills that all students will have some of, but none will have all. This framing sets up the expectation that there are a variety of forms of competence, and makes it easier to use status interventions like assigning competence.

2. *Group Worthy Tasks* are used to engage students in deep learning with problems that require multiple minds to meaningfully engage with. Because these deep tasks require multiple competences, it makes it easier for the productive ideas of many students to emerge.

3. *Group roles* support the delegation of authority to students and help them engage more equitably within small group settings.

4. *Assigning competence* is the practice of explicitly elevating the status of students perceived as low status by highlighting their meaningful contributions in public spaces.

This set of techniques was developed systematically over time through a series of studies spanning five decades.[18] The earliest studies were laboratory studies, which later transitioned to messier (and more authentic) classroom studies. Here I review a few studies documenting the impact of these strategies. One early study focused on the use of the multiple-ability treatment in a controlled experiment [295]. The researchers created two similar conditions but varied the framing of the task. It was framed to students in the comparison groups that reading ability was the primary skill they would need to succeed, whereas in the experimental groups, the tasks were framed so that the students believed they would need multiple skills unrelated to reading [295]. The researchers found that gaps between students perceived as high and low status related to the quantity and quality of their participation were significantly reduced in the experimental setting. In other words, students perceived as low status had significantly more opportunities to participate due to the status intervention.

The impact of these strategies to reduce status imbalances has also been studied in classroom settings [52]. In one study of 13 elementary math/science classrooms, it was found that using the multiple-ability treatment and assigning competence together could reduce status imbalances. The researchers used regression analysis, finding the rate at which teachers used status treatments was a significant predictor of talk from students perceived as low status ($\beta = 0.194, p < 0.05$).[19] In other words, the more frequently that instructors used the interventions, the more that students perceived as low status talked. Additionally, further analysis showed that the status treatments did not negatively impact participation from students perceived as high status.

In another early study, 28 elementary mathematics teachers were taught to use Complex Instruction, and their classrooms were observed [53]. The study was conducted across two years, with 15 classrooms in 10 schools during 1982–83, and 13 classrooms in 5 schools for a replication study in 1984–1985. The researchers were focused on efficacy of implementation; if instructors implemented Complex Instruction strategies with fidelity, it would result in greater collaboration between students. Indeed, there was a correlation between teacher instruction and student collaboration ($r = -0.429, p = 0.055$) in the original sample; ($r = -0.494, p = 0.51$) in the replication,[20] which were marginally significant.[21] Additionally, the authors found that the more that the students in a classroom talked and worked collaboratively, the better they performed on the standardized tests overall ($r = 0.72, p < 0.05$) in the original study; ($r = 0.52, p < 0.05$) in the replication study. Combining these two relationships provides a theoretical and empirically grounded model for how Complex Instruction impacts learning by modifying student participation, which consequently impacts their performance.

Over the years, the research team continued to implement and study Complex Instruction across settings [49, 156]. These reviews of Complex Instruction research (one book, and one more recent book chapter) document dozens of studies that took place in hundreds of classrooms. Across the studies, the authors provide various forms of evidence that Complex Instruction improved student collaboration, reasoning, and performance on outcome measures. Overall, the scope of the research program is impressive, and rare in equity research. Nonetheless, researchers have not conducted a meta-analysis or large-scale randomized controlled trial, so it is difficult to assess the exact impact of these techniques.

One possible pitfall with the Complex Instruction techniques is implementing them without attention to student social marker identities.[22] For example, prior work has showed that when the concept of status is taken up in a com-

[18] I already reviewed some of the development of the theory underlying these techniques in Section 2.2.2.

[19] Note that the mean rate of student participation was 4.37. The regression coefficient $\beta = 0.194$ means that each time an instructor used one of the status interventions (which were used infrequently), it corresponded to an increase in the rate of participation for students perceived as low status by 0.194. The exact definition of the participation rate was specific to the observation instrument used by the researchers, beyond the scope of this review.

[20] The correlation coefficients r near -0.5 indicate a moderate negative correlation between teacher direct instruction and student collaboration. In other words, the more that teachers guided the students, rather than delegating authority as they were taught to, the less that students collaborated.

[21] In inferential statistics, the cutoff of $p < 0.05$ is somewhat arbitrary, and so results near but just above this cutoff are considered marginally significant.

[22] Even in studies where these strategies are implemented in schools serving primarily minoritized students, it is still important to look at the variation in experiences by minoritized identities.

pletely colorblind way, it could actually reproduce greater inequities [173]. As others have argued, when Complex Instruction is taken up with specific attention to racial equity, it can be a powerful set of tools,

> [Complex Instruction is a powerful anti- racist pedagogy because it attends to dismantling hierarchies of competence that are steeped in race as a diffuse status characteristic. Let's break down a system that claims that young people with dark skin are not smart. We will rebuild our classroom communities with focus on their brilliance and interdependence among all students. (Lisa Jilk, personal communication, March 23, 2021; as quoted in [156]).

In my own work, I have found that these teaching techniques can be very effective for reducing racial and gender equity when they are paired with data showing racial and gender inequities. These data allow instructors to use the strategies with intentionally to support specific minoritized students in their classrooms. In other words, Complex Instruction strategies can be very powerful, especially when combined with the social marker specificity of EQUIP that disaggregates data along student identities like race, gender, and disability.

2.3.8 In Conclusion, to Lecture or Not?

As argued in the beginning the chapter, the answer to this question is a resounding yes, and.... Foundational research in educational psychology shows that active ways of engaging with material (e.g., practice testing, generating explanations) produces greater learning. These findings are corroborated by field studies in K–12 mathematics, as well as a meta-analysis of postsecondary STEM environments. These findings are all concordant with what has been learned by the field of education over the past 100 years and prominent theories of learning.

If pure lecture is not the most effective teaching method, why is it so prevalent? There are a variety of reasons, rooted in both personal and systemic factors. Today's faculty largely consists of people who found ways to succeed in primarily lectured-based environments, while others were filtered out. However, I would argue that most of us learned despite certain teaching practices, not because of them. Lecture also provides an alluring false sense of confidence. Research shows that even though students in pure lecture courses learn less, they perceive that they are learning more than their peers in an active engagement context [70]. In a lecture, the instructor does the hard work, making the content feel easy. An active classroom is much messier, and students struggle, make mistakes, and must work through repeated failures. Even though they are learning more through the process, it gives the (proper) impression that learning is hard work.

Systemic factors also play a major role in the sustained dominance of lecture-based environments. The academy is driven by an ideology of *publish or perish,* and as a result, publishing papers and procuring research grants are the primary external pressures driving faculty behavior at many institutions.[23] Moreover, learning to teach effectively in an active learning classroom is challenging—a lot more challenging than lecturing—and many of us don't have the time or privilege of receiving a formal education in education, although this is beginning to change. Also, as institutions are driven to increase efficiency, there is a push to pack more students into a smaller space, and this means that many of us teach in large lecture halls that really aren't designed to support active engagement. For all these reasons, it is hard to change the status quo, even for faculty who really are invested in learning to teach more effectively.

Despite all these factors, in undergraduate mathematics, the move away from pure lecture and the uptake of active engagement techniques has surged over the past decade. The growing support for these teaching methods is evident, for example, in the recent statement from the Conference Board of the Mathematical Sciences (CBMS). This statement—signed by the leadership of nearly every major mathematical organization in the US—called for "mathematics departments" to "invest time and resources to ensure that affective active learning is incorporated into post-secondary mathematics classrooms" [205, p. 1]. Shortly after this statement, the Mathematical Association of America released the *Instructional Practices Guide*, which is an extensive volume of active learning strategies that mathematics instructors can use [202]. As should be clear, there is growing momentum for these ideas in postsecondary mathematics. Although some of the content in this book overlaps the *IP Guide*, one of the biggest differences is the explicit focus on disrupting hierarchies and classroom inequities, whereas the *IP Guide* focuses on teaching more generally.

[23]Even graduate students who pursue a PhD primarily due to an interest in teaching are told that they must first succeed at a research and build out their publication record so that they can land a teaching job.

This exciting movement is not without caveats. Research strongly suggests that while active classrooms produce greater learning gains in the aggregate, they can also exacerbate inequities for minoritized students. Over the past eight years, my research team and I have been working to address this issue. We have developed an observation tool EQUIP and a set of professional development methods grounded in disaggregated student data that can reliably produce changes to patterns of inequity in student participation.

2.4 Reflection

Now that you have reached the end of this chapter, here are some questions to support your further reflection. These questions will be most effective if you consider them in collaboration with others, who will have their own and different responses.

1. Think back to the guiding question of the chapter: in what ways is research about teaching mathematics consistent with what I already know, and what are areas for productive growth and learning to improve as an instructor?
2. Write a letter to a younger version of yourself. Now that you've learned more about this research, what is something you would tell yourself as a teacher to help when you were just getting started?
3. Think about your experiences as a learner of mathematics. What has worked well for you, and what didn't work well for you? Why do you think that was the case? Compare and contrast your ideas with others around you, especially those who are not math faculty.
4. Looking at the references of this chapter, what is one reading that is new to you that you would like to read in its entirety?

3

Hierarchies in Mathematics

This chapter will provide you with a theoretical framework for understanding hierarchies in the classroom and empirical data documenting how these theories play out. Before you read this chapter, I'd like you to take a minute to write down some ideas you have about equity in mathematics, and where those ideas came from. Please take out a piece of paper (or use an electronic device) and take five minutes to write these things down. This extra processing time will make it easier for you to see what is new to you. In addition, this creates an opportunity for you to write down *new ideas* in this chapter as you encounter them. As you read the chapter, consider the reflection question: *in what ways is research about equity in mathematics and disrupting hierarchies consistent with what I already know, and what are areas for productive growth and learning to improve as an instructor?* Think specifically about ways that some students may have more privileges than others, and how your instruction can help account for those historical inequities.

3.1 Overview of Hierarchies

The learning cycle of instruction-practice-feedback is fundamentally correct, insofar that people learn through participation in disciplinary practices. However, this simplified view of learning is incomplete, as it does little to account for learning that happens in social contexts. Within social contexts, there are social dynamics, which almost certainly lead to hierarchies. As research shows [156], there is a causal link between status (or hierarchies), student participation, and student learning. For this reason, if we want to promote more learning for more students, we will need to disrupt the hierarchies that emerge in our classroom.

Because this book focuses on the US context, the exact hierarchies and systems of oppression that are most salient may differ than those in other parts of the world (e.g., a focus on learning in India would necessarily need to account for caste). Nonetheless, all contexts have hierarchies, and your local knowledge is valuable for understanding and disrupting them (e.g., consider the hierarchies between "pure" and "applied" mathematics). Because we live in a competitive (i.e., capitalist) society, hierarchies tend to emerge even in the "flattest" organizations. The pursuit of endless growth and conquest has fueled the genesis of oppressive systems like colonialism, racism, patriarchy, ableism, anti-Blackness, xenophobia, and so many others. Fundamentally, these systems exist to perpetuate inequities, to help individuals continue to accumulate unearned (i.e., stolen, inherited) wealth at the expense of others. The result is the creation of a society that is unequal and unjust, so that some may accumulate wealth beyond fathom while others starve.[1] This book is a guidebook for mathematicians and mathematics teachers who wish to divest from the unfortunate role that mathematics has played in perpetuating such a society. To be clear, I love mathematics and believe it is beautiful. I believe it can be a source of creativity, innovation, joy, and life. But the reality is that mathematics also can be—and historically has been—used towards much more pernicious ends [206].

[1] I find that this wealth accumulation in capitalism is akin to dragons in fantasy stories. The dragon collects a pile of gold simply to sit on it, while the villagers starve.

Given that life in modern society is increasingly organized around technology, broadening success in mathematics is often framed as an economic imperative [201]. According to this rhetoric, the United States (or any other nation, for that matter), should invest in technology to stay globally competitive. As a gatekeeper to STEM fields, success in mathematics is used as a key index for the efficacy of the education system. And conversely, a lack of success in mathematics is framed as a threat to the future prosperity of a nation [214]. Evidently, mathematics education initiatives are closely intertwined with capitalist and economic imperatives to generate more wealth. To be clear, the emphasis on disrupting hierarchies throughout this book is not just another ploy to further capitalist expansion but is a necessary step for the health and well-being of our students and society.

3.1.1 Mathematics as a Signifier of Inequity

Mathematics has a special status in society, which uniquely positions it to signify—and thus perpetuate—inequity. Commonly, mathematics is portrayed as rational, perfect, objective, and complete, unlike any other body of knowledge [80, 93]. The argument is that while the arts deal with matters of values and opinions, mathematics deals with matters of *fact*. The statement that "2+2=4" is taken as incontrovertible evidence of the perfection of mathematics, even though it is a trivial and misguided characterization of the discipline. While mathematics deals with degrees of certainty far beyond other disciplines, it is still imperfect, and influenced by human judgments, flaws, and biases [93, 104, 248].

This perfect view of mathematics has made it the de facto signifier of intelligence, through IQ tests, standardized exams in schools, or college admissions [6]. It has also made mathematics a tool for racism, as it has been used to justify slavery [14], eugenics [286], tracking [199], and other insidious endeavors [206]. These activities turn mathematics into a system of money laundering for oppressive ideologies. Tests of mathematical proficiency are used to track students into special education and gifted programs, in a seemingly objective way, even though the process is inextricably tied up with ableist racism in the schooling system, as students of color tend to be overrepresented and underserved by special education [3, 218]. Elite institutions of higher education are seen as serving the "best and the brightest," even though they often serve the richest and most privileged [41, 181]; consider that 14.5% of students at Ivy League colleges have parents who come from the top 1% of the income distribution, while only 13.5% of students come from the bottom 50% of the distribution [41]. At the same time, graduating from one of these institutions is used as an elite signifier of intelligence and success, rather than more accurately being seen as a signifier of immense economic privilege. And given historical legacies of slavery and eugenics in the US, such accumulated privileged is often (but not always) connected to racial inequities.

Perhaps more than any other discipline, mathematics is associated with testing, intelligence, and economic prosperity. As a signifier of intelligence, mathematics assessments play a unique role in student identity development. The dominant paradigm in mathematics is to assess students through timed exams in a high-stakes environment. Even though these tests are inauthentic and do not represent the practices of mathematicians, they have come to signify *school mathematics*, and therefore success on these timed exams is taken as a valid measure of one's mathematical ability. When a student receives a low grade on one of these tests, it plays into the narrative that some students are good at math while others are not, positioning the student as inherently bad at math. The individualized nature of assessments also communicates to students that mathematics is an individual and competitive enterprise, rather than a collaborative effort.

The interpretation of grades and their meaning is grounded in cultural frameworks. For example, Wu and Battey [320] conducted a 15-month ethnographic study of Asian American students to understand their experiences with the college application experience, vis-à-vis the expectations of their immigrant parents. One of the students in the study described the concept of "Asian failing," as a circumstance in which an Asian American student received only a A-, "because there is an expectation that Asians need to be able to get all As to stay on par with the other Asians in the school and community" [320, p. 591]. As this quote makes clear, Asian American students can be under intensive pressure to perform, and even the slightest deviation from perfect is seen as failure. In this way, grades are undeniably powerful in diminishing the self-confidence of students, even those who are ostensibly very successful at mathematics by conventional measures. Mathematics relies heavily on purportedly objective standardized examinations, even though the field of education still struggles to develop robust measures of achievement and learning. Nevertheless, standardized measures purport to provide an objective measure, while enriching testing companies [5].

Yet, these measures are strongly impacted by accumulated wealth, privilege, racism, ableism, and other factors that are nearly impossible to disentangle [5].[2] In this way, the testing industrial complex is a system of money laundering, because it transforms societal ills into "objective measures" that determine the futures of students, often perpetuating existing hierarchies. In other words, mathematics allows accumulated advantages to be packaged as merit and intelligence through the testing industrial complex. Despite efforts to use education as a tool for social change, educational systems strongly reinforce and bolster the oppressive status quo. Mathematics is at the heart of this process because it is broadly perceived as being fair even when it is being twisted to serve dubious ends. This is clearly a misuse of the beautiful discipline that mathematicians know and love.

3.1.2 Discourses

Hierarchies in mathematics are not only the accumulation and repackaging of privileged upbringings and identities. Hierarchies are also actively (re)produced within society and within classrooms. One of the most powerful tools for understanding how this works is a Discourse (a concept from poststructuralism). My own understanding of Discourses in mathematics education has largely been guided by the work of Niral Shah who has focused on the role of racial Discourses in mathematics [264, 265, 268]. A Discourse is a collection of symbols, signs, artifacts, and other cultural representations that work together to constitute the social world. Discourses constrain how individuals act, by creating a limited set of *subject positions* that they are permitted to occupy [290]. In this way, Discourses are mechanisms of power, because they define what individuals are permitted to do within a social context [87].[3]

The abstract concept of a Discourse can be better understood through an example. Consider how gender works in society. Gender divides the world into different categories, or subject positions (e.g., man, woman), and exerts power over individuals by dictating acceptable behaviors, from the clothes one wears to their career aspirations. This Discourse is also laden with narratives and expectations about what it means to be a *good man* or *good woman*, and overall, it privileges men within society as a part of the system of patriarchy. The dominant Discourse of gender in the US is a binary notion of gender (man vs. woman), but that Discourse is contested. For example, when individuals identify with and act in ways that create new subject positions (e.g., a non-binary person), it can shift the dominant Discourse. At the same time, attempting to shift a dominant Discourse can lead to backlash, as we see in recent legislative attempts to legally mandate the binary notion of only two genders.

There are also different ways to inhabit such a subject position (i.e., there is not only one way to be a non-binary person). There is not a single gender Discourse in society, but there are many, and they are historically and contextually bound. For example, queer communities have gender Discourses that are far more nuanced than mainstream society [254], and some cultures have entirely different Discourses with five or more genders [279]. It follows, that in understanding power in any social situation, it may be necessary to attend to multiple overlapping Discourses. Gender Discourses both create categorizations of gender (i.e., gender groups) and then exert power over the very groups that are created.

My personal understanding of Discourses has been informed by my lived experiences as a disabled and non-binary person. Dominant Discourses around ableism in the US position disabled people as less than their nondisabled peers. For example, throughout the COVID-19 pandemic, disabled lives have often been framed as disposable. In mathematics education, ableist Discourses relegate disabled people to the margins for either being too good at math, or not good enough at math [225]. Disabled people are also largely invisibilized in mathematics, with little recognition of the accomplishments of disabled mathematicians.[4] Belonging to a negatively stereotyped group in mathematics produces a variety of negative consequences (e.g., a reduced sense of belonging). Simultaneously, my white masculine presentation has shielded me from the impact of other negative discourses. To be clear, the impacts of ableism are not interchangeable with other systems of oppression (i.e., I wouldn't claim to understand the lived experiences of people who experience racism in mathematics on a regular basis, nor would I attempt to compare the two types of experiences.).

[2]For example, both racial [88] and gender [57] biases in standardized tests have been well-documented and have been a subject of litigation and legal concern for quite some time.

[3]It's important to distinguish between a Discourse in poststructuralism and the more colloquial usage of the term *discourse* which focuses on a classroom discussion. Throughout the work, I use the Capital "D" Discourse to signify that I'm drawing on poststructuralism.

[4]If you doubt this statement, I challenge you to come up with a list of successful disabled mathematicians, without using external resources. Can you come up with five? With ten?

3.1.3 Mathematical Discourses

In mathematics, there are a variety of Discourses that create different subject positions for students, such as those who are "high" or "low" students [157], or "students who can do math" vs. "students who cannot" [307]. These Discourses also tie into students' social identities, like "Asians are good at math" [320], "women are quiet and compliant" [159], or that "disabled students are deficient" [137]. Discourses both create social categorizations, and then exert power over the groups of students named (e.g., Asian students, women, or disabled students).[5] Relationally, they also impact other groups. For example, students of color who are not Asian are therefore positioned as not good at math [264]. Notably, these Discourses intersect students' myriad identities [44]. Discourses in mathematics provide a mechanism for hierarchally sorting students based on various social identities. For example, these Discourses position nondisabled White/Asian men as most capable at succeeding in mathematics. When someone has identities that differ from this reference point—especially multiple identities—they are less likely to be seen as capable.

Much of the research on Discourses in mathematics education has focused on race. Rhetorics such as the "achievement gap" position all racially minoritized students of color as less successful than their white peers. This fixation on comparison promotes a static, deficit view of racially minoritized students and their communities [102]. The way these logics operate, and how they impact students, can be understood through a lens of white supremacy [11, 149]. For example, Danny Martin's work shows how when Black children are framed as deficient, teachers describe lower expectations for Black learners in their classrooms [161]. Similarly, Discourses that Latinx students do not care about school or cannot speak English locate problems within Latinx learners, rather than societal injustices [97]. When such Discourses combine with Discourses around femininity, the impact is magnified to further invisibilize Latinx girls [268]. Overall, these Discourses locate problems within students and their communities, while obscuring the role of systems of oppression that perpetuate inequity. As this research makes clear, mathematics is not devoid of culture, but rather, is racialized, gendered, and so forth.

Underlying these Discourses is a view of mathematics as a discipline that is cold, logical, and competitive. Mathematical success is comprised of executing computations quickly and accurately. It is assumed that some people are innately good at math, while others are not. This success typically (but not exclusively) correlates to particular social identities. These logics have their roots in eugenecist and white supremacist movements that aim to classify certain races (i.e., the most "white") as superior and more intelligent than others [286]. Such conceptions of mathematics as ultralogical are often taken for granted without interrogating their oppressive roots. As we disrupt these common perceptions about mathematics, it provides a tool to disrupt the hierarchical categorization of different types of mathematics learners. Ultimately, we might aim to disrupt these very categorizations themselves, as we work towards mathematics education that is grounded in hope, joy, and community.

3.1.4 Disrupting Discourses

Another way to understand how Discourses manifest is through the concept of status [49]. As described previously, this refers to whose ideas have more weight in classroom discussions and who gets more opportunities to contribute. When we use instructional moves to treat status imbalances, we can intentionally disrupt dominant Discourses. In other words, when we co-create a classroom environment that highlights the brilliance of groups of students who are historically not positioned as brilliant in mathematics, we defy stereotypes and provide counterexamples to false classification of our students. To our benefit, there is a wealth of research and practical techniques that we can leverage to disrupt hierarchies in our classrooms, as briefly reviewed in Section 2.3.7.

In summary, you and your students are situated within a highly oppressive society, and unless you actively disrupt the hierarchies created by those systems of oppression, you may unintentionally reproduce them in your classroom. Trying to create an active learning environment without simultaneously disrupting these hierarchies likely means only a small subset of your class will get to actively engage. The hierarchies you notice in terms of who participates in your classes and who succeeds on your exams are not simply a matter of individual motivation, effort, or aptitude, but largely reflect our inequitable society. Some students have access to historically accumulated privileges, and others do not. Some face systemic and structural barriers that others do not. For your students who are not participating,

[5]To be clear, there is nothing natural about the classifications of "Asian students," "women," or "disabled students." These categories are all social constructions, and in a different system of Discourses, entirely different categorizations could exist instead.

don't assume why they are not participating. They may not have access to the material. They may be dealing with anxiety or trauma. They may have had a multitude of negative racialized and gendered experiences in previous math classrooms that have taught them mathematics classrooms are not a safe space. Your job is to create a safe and inviting environment that supports all your students—especially those at the margins—to be meaningfully involved.

I take as an axiom that all students are brilliant and capable of success. Rather than noticing what students can't do (i.e., focusing on their deficits), you'll need to develop a mindset of seeing what your students can do, so that you can build on their strengths [126, 210]. If you believe that some students are less capable than others, you are likely to (re)produce these beliefs through your actions. It takes a lifetime of unlearning oppressive Discourses that we're surrounded by every day in an oppressive society. Throughout this book we will work on practical tools that you can use to slowly start disrupting those larger societal hierarchies.

3.2 Material Consequences of Discourses

Discourses are not just stereotypes, but have material consequences. These consequences include low expectations, limited opportunities to participate, negative interpersonal interactions, stereotypes, and stereotype threats [137, 141, 159, 163, 280]. The way these mechanisms play out is not uniform across identities, and students who are marginalized across multiple dimensions of their identities (e.g., race and gender) face unique forms of oppression, as Maisie Gholson has highlighted in her work with Black girls [94]. In many ways, Discourses produce self-fulfilling prophecies. The presence of pernicious narratives positioning some students as not capable at mathematics thereby invokes various mechanisms of oppression, which inhibit the success of that very group.[6] For example, for some disabled students, the negative stigma associated with special education can often outweigh any purported benefits of extra support, thereby making disabled students worse off, and even producing the very notion of disabled [4].[7] I caution against playing "oppression Olympics," which pits minoritized communities against one another by focusing on who has it worse (e.g., when Asians are positioned as a model minority [265]); we need collective liberation to dismantle these forms of oppression. Here I outline some of the mechanisms through which Discourses impact students in more depth.

3.2.1 Participation Opportunities

A variety of studies demonstrate how participation is significantly related to learning [10, 53, 122, 238, 312]. Moreover, as elaborated previously, Discourses impact who gets to participate in classroom activities, and how they get to participate. Instructor biases—which are grounded in Discourses—impact who they call on and what types of questions they ask [166].[8] As a result, more privileged students are likely to be called on to answer questions more often, and they are often asked the higher-level mathematics questions.

In addition to biases, Discourses position some students as more capable at mathematics—and in the absence of an instructor intentionally mediating the discussion—these students tend to dominate classroom discussions. For example, masculine Discourses in mathematics produce a classroom context in which men exhibit louder, aggressive, and competitive behaviors, which in turn, means that they take up a disproportionate amount of talk time and learning opportunities [78, 228]. In small group settings, student status mediates who gets more talk time and which students are trusted as authorities [138, 139]. Together, all these mechanisms provide more participation—and thus learning—opportunities to students who are positioned as capable at mathematics through dominant Discourses. In Section 3.3, I discuss some of my own studies of classroom participation in depth.

[6] As described previously, Discourses actually produce these classifications. For example, in mathematics the division between "high" and "low" students creates an arbitrary system of categorization, and because some students are thereby framed as deficient, they are negatively impacted by low expectations and other mechanisms. These concepts of "high" and "low" are also almost inevitably intertwined with identities such as race, gender, and disability.

[7] Simultaneously, privileged parents are able to game the systems of accommodations to provide further benefits to their already privileged (and often nondisabled) children [68, 69]. This provides another example of how policies and structures that are intended to reduce inequity can further (re)produce it.

[8] In the Section 3.3 I provide a much more extensive review of student participation opportunities.

3.2.2 Belonging, Identity, and Stereotypes

Research shows a significant relationship between belonging and persistence in college [100, 108, 178]. For example, consider a study that used a nationally representative survey of first-year college students ($N = 23,750$) with a two-year follow up [100]. The study used multiple regression to consider sense of belonging alongside a variety of other characteristics and demographic variables and found that belonging significantly predicted persistence. They also found that students reporting very low levels of belonging ("strongly disagreed") had significantly lower levels of persistence (14% lower compared to "strongly agreed"). Notably, these low levels of belonging were significantly more likely for racially minoritized students compared to White peers ($\chi^2 = 15.47, p < 0.001$)[9] and first-generation students compared to continuing generation peers ($\chi^2 = 8.00, p = 0.005$). Similar findings exist for STEM persistence, where students who feel like they don't belong are less likely to continue pursuing a STEM-intensive future, even if they are otherwise well-positioned to succeed [262, 296].

In mathematics, minoritized students must contend with problematic stereotypes and narratives describing them as math learners – or more precisely, not as math learners [140, 141, 285]. For example, if a Black student were to walk into a calculus classroom full of primarily white students, it often invokes racial stereotypes about math ability and who belongs, based on who is in the room [141]. This stereotype then creates another barrier for that student—who may be implied not to belong—regardless of that student's prior math background or experiences. The student may experience the *imposter syndrome,* a concept which helps describes the additional pressures that result from negative stereotyping [47, 209]. Overall, being positioned as the *other* reduces access to meaningful learning opportunities and the types of engagement that could support success. Learning environments communicate belonging in many ways, including subtle context cues such as the decorations on the wall, which can significantly reduce student interest to participate in a discipline as a result [40].

3.2.3 Microaggressions

Microaggressions are toxic interpersonal interactions that are grounded in students' social marker identities [289]. These include questioning someone's belonging, questioning someone's intelligence, devaluing someone's identities, and so forth, based on that person's social marker identities (e.g., saying "wow, you're in this calculus class!" with a surprised tone to a Black student). Because STEM classrooms (including mathematics) are highly stereotyped, and microaggressions arise commonly when students are interacting with peers, they commonly arise in mathematics classrooms. This has been documented across a variety of social marker identities, such as race [146, 169], gender [78, 260], disability [75, 151], sexuality and queer identity [59], and so forth. Microaggressions may also be tied to specific aspects of mathematics discourse, like saying "this is trivial," in a way that inhibits belonging, especially for students who are already minoritized [287, 38].

Microaggressions create a hostile and stressful environment for students from minoritized groups. This produces a variety of measurable impacts on students, including reduced self-efficacy and self-esteem, otherness, isolation, and stereotype threat [303]. These events can become so toxic that they push minoritized students to change classes, majors, or even drop out of the university [275]. Rather than focusing on learning, students must focus their effort into managing and avoiding these hostile incidents. Additionally, students may be hesitant to participate because they want to avoid microaggressions from peers or their instructor.

3.2.4 Low Expectations

Teacher expectations impact student performance. For example, a study of gendered expectations in elementary mathematics found that teachers systematically underestimate the abilities of young girls, and these misperceptions could account for approximately 50% of observed performance differences between boys and girls [242]. Similarly, Black students often must contend with low teacher expectations in a way that limits their opportunities to engage and be seen as successful [161]. Expectations are communicated in a variety of ways from subtle to overt: from who gets called on to participate to how teachers explicitly praise or reprimand students. The impact of low expectations can

[9] χ^2 is a statistical technique often used to compare categorial data between groups. The specifics of this statistical method aren't important here, but merely that there was a significant relationship found.

also be implicit or unconscious, insofar that instructors don't even realize it, but they are acting in accordance with stereotypes in a way that actually creates those inequities [278].

Biased expectations also impact additional opportunities that may be afforded to students. This phenomenon was demonstrated strongly in a randomized study of science faculty's biases [180]. In this study, faculty members were shown identical application materials for a lab manager position, except that the name of the student was randomly assigned to be a man or woman (non-binary students were not included). Faculty consistently rated the men as more competent and hirable than the identical woman candidate. Similarly, the men received significantly higher starting salaries ($30,000 vs $26,500 for women) and more career mentoring. These biases against women in science were exhibited by faculty regardless of their gender. This study provides a strong demonstration of how negative stereotypes (in this case, women in science) produce measurable, material consequences that enhance the career opportunities for some students over others.

3.2.5 Stereotype Threats

Dominant Discourses also have a strong impact on student success on assessments. This mechanism has been described by Claude Steele as a *stereotype threat* [280]. A stereotype threat occurs when a negatively stereotyped student is put in a high stress situation (e.g., a high-stakes exam), and their performance in that situation has the potential to reinforce that negative stereotype (e.g., confirming the false notion that women or minoritized students of color are not good at math). The damaging impacts of stereotype threat are well-documented and can account for many disparities in mathematics test performance [188, 277]. Meta-analyses show a small effect size ($d = 0.22$) for women in math [86], and medium effect size ($d = 0.52$) for Black and Latinx people on tests of intellect [182]. These findings are especially alarming, given that high-stakes timed exams are the de facto standard in mathematics, and other types of assessment are typically discredited as not rigorous. However, recognizing the inherent unfairness of high-stakes tests (through both stereotype threats and a variety of access issues), alternative practices should be used to mitigate the further perpetuation of inequities.

3.2.6 Overall Impacts

As the variety of mechanisms elaborated above show, Discourses that position some students as capable at math and others as not capable are one of the very tools that create inequities. When students are positioned negatively within these Discourses, it limits their access to competent subject positions within the Discourse. The concrete mechanisms such as low expectations, microaggressions, stereotype threats, and fewer participation opportunities all work in ways that undermine learning and belonging. Regardless of whether an individual instructor or a student believes these Discourses, the very existence of such Discourses produces these effects. As a result, instructors must work to intentionally disrupt these Discourses, or they will only reproduce hierarchy in their classroom.

3.3 Equity Analytics

While Discourses provide us with a theoretical tool to understand why hierarchies emerge, patterns of student participation provide empirical evidence of the various hierarchies that are present in the classroom. Over the past eight years, I have been co-developing the equity analytics approach to understanding these hierarchies with Niral Shah [230]. In common usage, equity refers to issues of "justice" and "fairness," meaning that all students should get what they "need" or "deserve" as learners. Unfortunately, few people can agree on what fair should look like, much less develop a methodology to operationalize it for mathematics teaching.

To make progress despite these issues, we draw upon the concept of equality as a waypoint towards equity [259]. Equality is a situation in which all learners receive the same thing. This is easy to measure, but at the same time, it is clearly unfair, because different students have different needs. Given persistent historical inequities and mechanisms of oppression, we argue that in any classroom where minoritized students receive fewer opportunities to participate relative to their non-minoritized peers, that would signify inequity. In other words, we need to intentionally work to repair historical (and present) harm (i.e., a reparations view of equity). Although this still does not provide an adequate definition of equity, it is more than sufficient for improving the status quo. In over eight years we've observed hundreds of classrooms, and in every single one of them, minoritized students received less than an equal share of participation opportunities. One day we hope this is no longer true.

To operationalize the principles of equity analytics, we have developed the EQUIP observation tool to create a methodology for tracking patterns of student participation [230]. In this section, I provide an overview of the tool, the scope of work to date, results from select empirical studies, and an overview of the professional development process.

3.3.1 How EQUIP Works

EQUIP works by tracking student participation in classroom activities. In most studies we have focused on verbal talk because it is the most public and easiest to capture form of participation. However, in some studies we have also captured a variety of participation modalities, including written, gesture, ASL, manipulatives, and so forth. In short, anytime that students engage in public activities that involve expressing mathematical ideas, they can be captured with EQUIP, because EQUIP is a customizable methodology.

The basic unit of analysis for EQUIP is a student contribution. For verbal contributions, we analyze all talk from a single student that is not interrupted by another student. (A single contribution could involve back-and-forth between a single student and the instructor. This allows us to account for situations where a teacher follows-up with students to elicit deeper thinking, rather than simply coding the initial, shallower, contribution.) As soon as a new student participates, it creates a new contribution. By segmenting the observation by contributions, EQUIP can answer the *who* question about what happens in the classroom, by uncovering how different students have different experiences. In essence, all teacher and student behaviors that are coded are attached to specific contributions, which allows contributions to be aggregated or disaggregated across students and groups.

For each contribution, coding takes place across a variety of discourse dimensions. Discourse dimensions are features of participation such as the length of the contribution, the type of contribution, the types of questions a teacher asks, how a teacher responds to a student, and so forth. Because EQUIP is customizable, the exact dimensions used in any study may differ. Nonetheless, across a wide variety of studies, dimensions such as length of contribution, teacher question type, and whether students are called on (or spontaneously contribute) have provided useful information about classroom interactions. For the sake of simplicity, most of the times I talk about EQUIP in this book I focus just on amount of participation, not the quality of participation. Suffice to say, these extra levels of detail allow us to uncover more subtle patterns of inequity, beyond the larger, more obvious patterns. That deeper level of granularity can be found by reading one of the many journal publications my team has produced about EQUIP [230, 238].

To generate analytics at multiple levels of the classroom, EQUIP draws upon student demographics. Student demographics are typically gathered through a survey. The conceptualization of demographics is broad – essentially any categorical variable that can be put on a survey could be added as a demographic. This often includes race and gender, but could be disability, language, major, socioeconomic status, group membership, religion, and so forth. Which demographics are used depends on the context and the needs of particular instructors. Because each contribution is coded at the student level, once demographic information is added, contributions can be aggregated across any demographic variable or cross section of variables (e.g., race and gender). EQUIP can also provide information about the whole class, or individual students. This level of granularity at multiple sizes has made EQUIP a very powerful tool for research and for generating data that meaningfully change practice.

Once all these data are gathered, an observer is ready to code with EQUIP. Typically, an observer will video record a class session (or virtual meeting) to code it later. However, real time coding is also possible under some circumstances. One benefit of coding videos is that it allows an instructor in one location to record their teaching, upload the video to a coach in another setting, and debrief virtually. A free web app (`https://www.equip.ninja`) has been developed to facilitate the coding and debrief process. After coding, the EQUIP app can also be used to generate data analytics to be shared with instructors for professional development purposes.

One important metric that we have developed to understand inequities in participation is the *average contributions* of a particular demographic group (which we have also referred to as a *participation rate*, see [228]). This is a metric we developed to account for different sizes of different groups in the classroom. Suppose we had a classroom with 4 Black students and 20 Latinx students, and Black students contributed 8 times, while Latinx students contributed 16 times. While in the aggregate, the group of Latinx students has participated more, on average, Black students contributed more than twice as often. Although this example is simple, in a real classroom situation, we may be dealing with 5-6 racial groups, as well as intersections of other identities. Coordinating the quantity of participation with demographic group sizes quickly becomes a nontrivial task.

Thus, to account for the different sizes of demographic groups, we introduced the metric of average contributions for a group of students. We calculate this by dividing the overall number of contributions for each group by the number of students in the group. As a result, we can understand student contributions with respect to the demographics of the class. Now, if there were 8 contributions from 4 Black students, this would mean the average contributions were 2 for Black students (i.e., on average, Black students contributed twice). In contrast, the 16 contributions from 20 Latinx students would mean the average contributions for Latinx students were only 0.8 (i.e., on average, Latinx students contributed 0.8 times, or less than once).

3.3.2 Scope of EQUIP Activities

The equity analytics approach and EQUIP tool have been built up slowly and methodically over eight years. During the period from 2014–2022, my research team has published on the approach extensively, resulting in 16 journal articles published and in press. Other researchers have also taken up the tool, disseminating their work at conferences and in journal publications. To date, there are over 2500 users of the EQUIP app. While it isn't methodologically feasible to aggregate results from all this work directly, over nearly a decade my team has met many others using the approach to discuss findings and triangulate our understandings. This has provided insight into what happens in mathematics classrooms across all types of settings. While some of these findings haven't been formally published yet, I do my best to share from this work throughout the book.

My team has been involved in long-term professional development work that has spanned K–12 teachers and faculty (nearly 100 participants in total). Higher education settings have ranged from community colleges, four-year universities, PhD granting institutions, elite liberal arts colleges, and so forth. Instructors have taught in disciplines as widely ranging as mathematics, physics, computer science, public health, journalism, linguistics, counseling, anthropology, music, and so forth. While each discipline is a bit different, this wide range of experiences has truly demonstrated the flexibility and reach of our approach (with a bit of customization for the local context). While we have worked in a variety of disciplines, mathematics is the primary domain that the work has been situated in. Typical engagements with instructors have lasted from one to five semesters total. Learning to disrupt hierarchies and teach equitably is a lifelong process, and the more time instructors spend in the program, the more they enhance their learning.

3.3.3 Empirical Studies

One arm of the EQUIP research has focused on understanding typical participation patterns in mathematics classrooms. While there is quite an extensive body of mathematics education research that focuses on inequities through rich qualitative studies and uses ethnographic methods, there are inherent limitations to the scalability and comparability of such approaches. For these reasons, my team and I developed EQUIP as a quantitative approach that could complement—not replace—such work by providing metrics that could be used across settings.

Here I review some of the key findings from that work. These studies have taken place in mathematics classrooms across elementary schools [230], middle schools [238, 268], and undergraduate classrooms [78, 226, 231, 228]. We have also done work in physics [266] and even a mix of STEM and non-STEM disciplines [232]. Although each setting has its own nuances, in general, what we have found is that hierarchies in mathematics seem to emerge at a very young age, and they persist throughout students' mathematical journeys. We also find that once teachers become aware of these patterns using data, they can make considerable changes to their teaching practice that shift those patterns (I review some of those findings in a Section 3.3.4).

3.3.3.1 Findings about Gender

Our most consistent finding across studies is widespread gender inequity. Every study that we have conducted in mathematics classrooms with a focus on gender has illuminated some sort of inequity in favor of boys/men [78, 231, 232, 230, 268, 266, 267]. Our research hasn't been limited to a binary view of gender, and in fact, drawing attention to the experiences of nonbinary and genderqueer students is one useful advantage of the equity analytics approach for professional development. Yet, in this section, where I am making statistical generalizations, I focus only on a binary view of gender, given the limitations of small numbers of nonbinary students in our samples. Our findings around gender corroborate earlier research across disciplines, which focused on the ways in which girls receive fewer opportunities to participate—and thus learn—within K–12 classrooms [251].

The largest of these three studies focused on participation in middle school classrooms in a racially diverse urban school district [238]. We performed a secondary analysis of existing data, which came from a larger professional development study working across four districts. The district we chose to analyze with EQUIP was the most instructionally advanced of the four districts (according to a variety of metrics developed on the initial project). Our sample included 100 middle school mathematics classrooms across the district, which were studied over four years. Each classroom was observed across a single unit of instruction, which consisted of two video recorded lessons (approximately 100 minutes of class time total). We found that the average contributions for boys was 2.10, which was significantly greater than the 1.72 average contributions for girls [238]. The differences were most significant in cases where boys were *not called on*, that is, when they made spontaneous contributions without instructor intervention. As discussed in Section 3.3.3.2, we also studied how these results intersected race. As far as classroom observation studies go, this one was relatively large in scope, especially considering the detailed coding that goes into using EQUIP. These results show how the impacts of gender Discourses are wide reaching across a variety of mathematical settings.

The second study, which I introduced briefly in Section 2.3.5, focused on inquiry-oriented instruction in upper division undergraduate mathematics [228]. This study was organized around a reasonably robust and ongoing professional development program that was provided to mathematics instructors through a summer workshop and throughout the summer. While students in the inquiry-oriented classes learned more in the aggregate, the learning benefits were not equal for men and women, and as a result, the inquiry classes had significant gender inequities. To understand these inequities, we analyzed classroom participation by gender. Because this was another analysis of secondary data, we couldn't directly connect individual student participation to individual student outcomes. However, we could study gendered participation in the aggregate, and relate it to gendered differences in outcomes in the aggregate. In total, we analyzed 20 classrooms (from Abstract Algebra and Differential Equations), each of which were observed for a single unit of approximately 145 minutes.[10]

Overall, we found that the average contributions for men (6.0) were higher than women (4.0). Most notably, the average contributions differed by classroom, and these could be connected to differences in outcomes. We constructed a weighted regression with the following equation:[11]

$$\text{(Gendered Performance Difference)} = -0.59 + 0.09 \times \text{(Women's Average Contributions)}$$

This regression equation shows that as women's average contributions increased, the gendered performance difference was reduced. The intercept of -0.59 indicates that with no participation from women, the performance difference was just over half of a standard deviation in favor of men. The 0.09 indicates that if women had approximately 11 contributions on average ($0.09 \times 11 \approx 1$), it would reduce the gendered performance difference by one standard deviation. In this study, the average classroom had about six women, so this would amount to 66 contributions total over the course of the 145-min unit. Given that the total contributions in these classrooms went up to the 200s, this is certainly feasible.[12]

The third study was smaller in scope, focused on a single undergraduate geometry course [78]. In that study, five cameras were used. Four cameras captured small-group interactions, while the fifth camera captured whole class discussions. This setup allowed us to compare participation in small-group and whole-class settings. Moreover, because the group cameras continued running during the whole-class discussion, we could continue to capture side talk during

[10] While the average unit length was 145 minutes, there was variation in the length of the observed unit for any given teacher. To allow us to compare across classrooms, all results were scaled to a 145-minute unit size.

[11] Our analyses focused on women's participation, because it was significantly correlated with their performance ($r = .54, p = .01$), whereas men's participation was not correlated with their performance. Detailed methods can be found in the original manuscript, and are beyond the scope of this short review [228].

[12] To be clear, this study did not have an experimental design that would allow us to establish causality. In other words, it doesn't allow one to claim that increasing women's participation would automatically increase performance in this study. Nonetheless, such a conjecture would be consistent with what is broadly known about learning and is not a far stretch.

the discussion. The gender demographics of the class were relatively balanced: 9 women and 7 men in small groups observed, out of 14 women and 10 men in the class overall.

As one would expect, there were far more contributions during small group work: 105 average contributions for women, and 111 average contributions for men. The large number of contributions is due to the free-flowing nature of small group discussion. The whole-class discussions were also interactive but still featured far fewer contributions. Most notably, there was a large disparity in women's participation when we compared public and private side talk. The results are given in Figure 3.1.

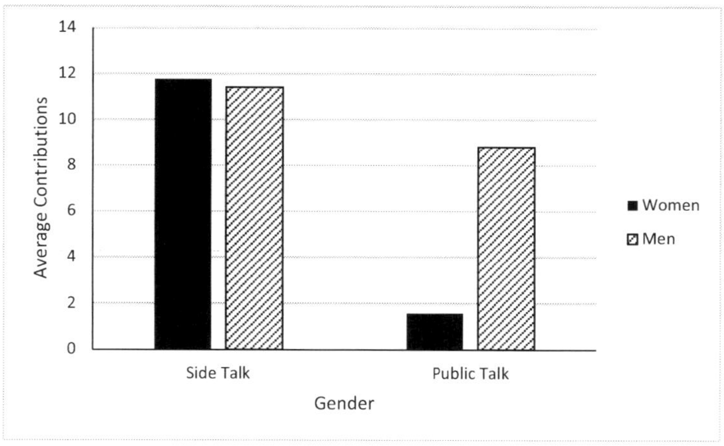

Figure 3.1. Private side talk and public talk by gender in an undergraduate geometry course.

Much like the small group interactions, contributions were relatively balanced between men and women for side talk. However, there was a striking difference in public talk. Women only had 1.6 contributions on average, compared to 8.8 for men. This indicates that women were engaged during the whole class discussions, they engaged privately at their tables, rather than publicly in front of the whole class.

To understand these glaring differences, we looked deeper into when the instructor was soliciting participation (17 instances analyzed). In three instances, the instructor called on a student by name, and only chose men. In 11 instances, the instructor called on the first student to raise their hand, and in 10 of 11 cases, it was a man. In the other 3 cases, the instructor called on a group, and in 2 out of 3 of those instances, a man spoke up first. Crucially, during the very same interactions in which men dominated the discussion, we found that women had already solved the problem but their participation was not made public (for more detail, see [78]).

At first glance, small group work seems to provide a more gender-equitable alternative to the male-dominated whole-class discussions. However, our analyses also revealed alarmingly sexist interactions in small groups. The first episode was drawn from a small group interaction in which students were using axioms to create a geometric proof. In this episode, a student Isabella was offering ideas for the proof, which were ignored by Raul, who disagreed, and ultimately responded with sexist tropes to devalue her mathematical thinking [78].

1. Isabella: Do this one.
2. Raul: Huh?

   ...

7. Carla: He does not listen to girls.
8. Raul: Huh?
9. Carla: Sexist, too.
10. Raul: Like I said before—what'd I say in the other class?
11. Isabella: What?
12. Raul: It's hard to hear her when she's in the kitchen. These aren't collinear.

Throughout this episode, Raul twice pretends not to hear the ideas of the women in his group (Lines 2 and 8, where he says "huh?"). After ignoring the women in the group twice, he invokes the sexist narrative about women in the kitchen to justify his behavior of ignoring the women. This type of blatant sexism is extremely harmful, as it invokes misogynist tropes in a way that intersect masculine narratives of mathematics to create a toxic and unwelcoming environment for women. Interactions of this sort were observed throughout the semester in this study. Later in the same class session described above, students had the following interaction:

1. Fabrice: Dude, that pen is not even straight. At all. Yeah, like that, there you go.
2. Carla: Axiom one, "any two distinct points are incident with one line."
3. Raul: It has to be with this line. God, you are silly.
4. Isabella: I like silly better than stupid. I know that you are trying to say the same thing.

   ...

14. Fabrice: You just have a bad perspective from where you are sitting.
15. Raul: Hey, fix your perspective.
16. Carla: I will. By removing two men from the table. I think axioms one, three and seven.

Again, we see Raul making blatantly sexist comments to invalidate the ideas of the women in his group. After Carla's suggestion of Axiom 1, Raul disagrees and says "God, you are silly." This type of statement is extremely problematic. Raul again is invoking a sexist idea that "women are silly" or that they cannot be serious in a mathematics classroom. In this case, Raul chooses to insult and denigrate the women in his group, rather than grappling with their ideas. These toxic interactions contributed to the environment that was overly hostile towards women, and we hypothesize played a major role in silencing their contributions during whole-class discussions.

3.3.3.2 Findings about Race

Our research has also documented racial inequities across a wide a variety of classroom settings [231, 232, 229, 230, 268, 266]. However, these results are less straightforward to summarize, given differences in the racial compositions of classrooms and particular racial dynamics at play in each setting. Nonetheless, overall, we find that white students are well-represented across settings with ample opportunities to talk, which mirrors the findings with gender, creating a race-gender hierarchy that favors white men (and some, but not all, groups of Asian men). Here I discuss a few studies to explain these results.

Depending on the study, we collect racial demographic information in different ways. In a K–12 setting, schools often have already collected this information from students. This is beneficial but it constrains our analyses to predefined racial categories. In a college setting, typically we would send out a survey to students and allow them to self-identify. This provides more flexibility for multi-racial students to describe their identities.

The first EQUIP study took place in a summer program for elementary school students who were entering fifth grade in the fall [230]. This course was taught by a seasoned educator with an explicit focus on equity. Student participation in whole class discussions was analyzed over two weeks (3 hours of instruction per day over 10 days). The class consisted of a mix of Black, Latinx, and White students. We found that Black students participated nearly proportionally to their demographic representation in the class. At the same time, we found that White students had the most opportunities to participate with higher-level explanatory *why* contributions and extended contributions of multiple sentences at a time. We found that Latinx students had significantly fewer opportunities to participate than their Black and White peers. These findings highlight some of the subtle ways that racial inequities manifest in classrooms even of the most seasoned educators who are working towards equity.

Returning to a previous study of 100 middle school mathematics classrooms, we explored gender, race, and intersections of the identities [238]. In this study, we found that White and Black students (1.80 and 1.92 contributions on average, respectively) had significantly more contributions than Asian/PI[13] and Latinx students (1.17 and 1.14 contributions on average, respectively).[14] To interpret these numbers, this means that on average, a white student would participate 58% more than a Latinx student in the observed classrooms in this district. These findings also intersected gender categories, with Black and White boys having the highest levels of participation, Black and White Girls having the next highest levels, and Latinx and Asian/PI students having the fewest opportunities to participate. It's noteworthy that this study took place in the context of a district that was engaged in sustained professional development to mitigate inequities for Black learners for many years. While the findings don't necessarily indicate equity for Black students, they did show some promise. Simultaneously, they show how other racially minoritized groups had limited opportunities to participate in these classrooms.

Next, I consider a study of three mathematics faculty members teaching undergraduate mathematics [231]. As an intervention study, there was no rigorous reporting of trends by race in the publication. Nonetheless, we note that all instructors had racial inequities in their classrooms. For one instructor, there were three Filipino men who were consistently marginalized in whole-class discussions, who became a focus for intervention. This is consistent with the experiences of another mathematics faculty member in another study who participated in a learning community [227]. While only 36% of that class was white, 51% of contributions came from white students. These contributions could be largely traced to three vocal white students (two men, one woman) who dominated class discussions. These findings are largely consistent with our other (unpublished) observations of mathematics faculty members in intervention work, in which white students are generally dominant, and those trends of dominance can often be traced to a few overly vocal students.

3.3.3.3 Other Identities

While research with EQUIP has focused primarily on race and gender, the tool is customizable, and we have also looked at a wide variety of other social markers. Some of these identities include disability, socioeconomic status, first-generation status, major, religion, and small group membership. The work with some of these social markers has not yet been published or took place in work that focused on professional development without research. Thus, we refrain from making any strong claims about how these identities play out based on our observations. Nonetheless, anecdotally, we observe that these work just as one would expect in classrooms. The more marginalized a student is, the fewer opportunities to participate they are afforded.

3.3.4 Professional Development

The other arm of EQUIP research has focused on professional development. Having conducted multiple studies that empirically document inequities in participation across settings, we recognized the urgent need to move beyond documenting a problem to helping to address the problem. This has led us to engage in instructor professional development. A key underlying principle and benefit to EQUIP-based professional development is the concept of social marker specificity. Social marker specificity means that any data generated in service of equity needs to be disaggregated along social marker identities (i.e., in order to improve racial equity for Black students, data collected should speak directly to the experiences of Black students, not just equity in general).

Most professional development approaches—and even classroom observations writ large—do not disaggregate data, and therefore, speak primarily to the classroom overall, while obscuring the different experiences that students with different social marker identities have. This can lead to conundrums, for example, when a professional development program is ostensibly about improving racial equity, but the data and processes involved only focus on race indirectly [173]. In contrast, we believe that addressing racial inequities requires noticing, naming, and disrupting racialized phenomena directly.[15] Because such phenomena can be subtle, having a tool to make them more visible and actionable

[13]The Asian populations in this study were primarily from Hmong refugee communities. It is important to recognize that Asians are not monolithic and different groups have different experiences despite an overarching narrative that all Asians are successful at mathematics all the time.

[14]An important caveat is that these studies did not necessarily have statistical power to investigate inequities across all racialized groups in each setting. For example, we found evidence of considerable marginalization of Native students in this study, but we couldn't explore it statistically due to small sample sizes.

[15]In this sentence, you could replace race with gender, disability, sexuality, or any other minoritized social marker identity that your work is attending to.

is extremely effective. Because EQUIP provides disaggregated data, it puts a focus on social marker identities front and center, which is key for providing meaningful data, and for producing lasting changes in how instructors think about classroom equity.

The basic model for using the EQUIP tool involves creating a small learning community of 3-5 faculty who engage in several reflection cycles in a semester with the support of a coach and coding team. This process allows faculty to engage in reflection and iterative improvements of practice. Typically, the goal is to have 3-4 cycles in a single semester to support meaningful change to instructional practice, so participants meet on roughly a monthly basis. The first meeting often happens about 5 weeks into the semester, to allow for intake interviews and setting up classrooms, so in practice, the debrief meetings might be about 3 weeks apart (with 4 meetings in a semester). Each cycle has four main components: 1) observation, 2) coding, 3) feedback, and 4) a debrief meeting.

The observation phase typically involves recording a class session of a participating faculty member. If a single camera is used, it is set up in the front of the room facing students, so that student participation can be captured. This recording is then uploaded to the professional development team. In the coding phase, a designated coder will code student contributions along some number of dimensions that were determined in agreement with the participating faculty member. Usually, the same coder will observe the same faculty member throughout the whole semester to help build greater familiarity with the classroom and to help notice subtle shifts in practice that may or may not be captured by EQUIP. Third, the coder generates a feedback report describing patterns of participation and inequity that were observed. The feedback report disaggregates individual and group-level patterns, includes qualitative observations, and provides some suggestions for changes to practice. The final step of an observation cycle, the debrief meeting, is an opportunity for the coder, coach, and team of faculty members to get together to discuss the results of their observations and next steps forward. This allows instructors to receive feedback and discuss with their coaches, and crucially, faculty can learn from each other.

The reflection process is iterative, so after a cycle finishes, each instructor sets a goal for changing the participation patterns in their class session and determines specific instructional moves that they might use. In the next observation, the coder can see to what extent the practices were used effectively and measure the impact on student patterns. By iterating these changes multiple times over a semester, instructors make meaningful changes to their practice. Our empirical work has documented impressive changes to practice in just a single semester (five hours of meetings total). These changes are further enhanced when an instructor participates for multiple semesters. As a research team, we have been able to work with many instructors and empirically measure when certain teaching moves impact participation patterns. Many of those same strategies are provided to the reader throughout this book. To illustrate the impact of professional development, I provide examples of two instructors who went through the program and the impact on their teaching. Pseudonyms are used to protect the identities of these instructors.

3.3.4.1 Anne

Anne was a white woman and a full professor of mathematics who had a successful career at an elite university. She was also known within the community as a strong advocate for women in mathematics, and served as a role model for women, because she herself was a highly successful woman in mathematics. As an instructor she liked to use real world examples—like how to develop cancer drugs—to build student interest in the subject. She found that students enjoyed the examples that helped them see mathematics as useful and impactful. As an instructor she tries to remove barriers that prevent students from reaching their goals. She tries to remove judgment and support confidence building. She doesn't use timed tests and allows for test corrections; her overall philosophy is to support growth. She was also very aware of racial and gender stereotypes in math and had experiences growing up of being told all the things that women can't do. She was determined to overcome such stereotypes in her own life.

The first semester she participated in the EQUIP learning community, she was teaching Linear Algebra. Her class had 29 students, with a gender breakdown of 17 men (59%), 11 women (38%), and 1 student with unknown gender. By race, she had 14 White students, 11 Asian students, and 4 students with unknown race. The first time her class was observed, there were 16 student contributions coded. These came from 12 different students (41% of the class). Of these contributions, 12 of them came from men, and only 4 were from women. Overall, no individual students dominated. In the first debrief meeting, Anne commented on this:

> I really want to address this business of the women not speaking. It bothers me. It just doesn't go away. They have a woman standing up in front of them, and they talk to me outside of class, but it's so hard, the social structures are so entrenched.

As this comment makes clear, Anne wasn't necessarily surprised by the initial data that she received, but it did confirm her prior experiences, and she was frustrated by it. Over the course of the semester, she started working with six new teaching strategies that were discussed in her learning community. Most of these strategies involved ways to bring different students into the conversation. Here are some of those strategies:

1. Discussing with students the importance of participation and creating activities to support varied forms of participation (e.g., moving rows and columns in Linear Algebra).
2. Waiting for multiple students to raise their hands before participating.
3. Implementing a "two voices rule" where students had to wait for two other students to participate before they participated again.
4. Explicitly asking to hear from students who hadn't spoken yet.
5. Carefully selecting students to share out after group work.
6. Turn-and-talk when few/no students were ready to answer a question/discuss a topic.

These strategies allowed Anne to be more intentional about who she brought into her classroom discussions. Through using these types of instructional moves, she was able to include women into the discussion who had valuable ideas but may not have otherwise contributed spontaneously. Over the course of the semester, Anne was observed four times. Here I focus on the gender-based trends, in specific (see Figure 3.2). Each of these observations was spread out by approximately 3–4 weeks.

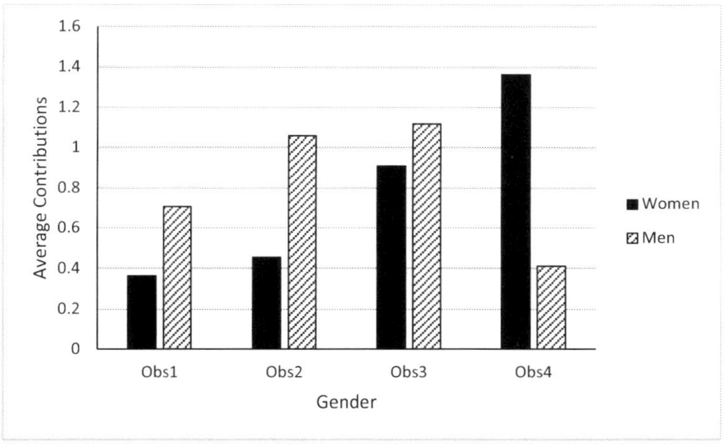

Figure 3.2. Average contributions by gender in Anne's class over the course of the semester.

As Figure 3.2 shows, as Anne became more adept with the specific teaching strategies that she used in response to the data analytics she received, she was able to intentionally bring students who otherwise weren't participating into the discussion. This allowed her to create a more engaging environment for students in general, and especially the women who were otherwise being marginalized. In reflection on the experience, she noted:

> This project has made me so aware about participation, which is awesome. . . I think the strategies that you all have suggested are really effective, and I need to keep working on them.

As this example shows, having lived experiences of discrimination as a woman in mathematics, and having an intentional goal of supporting women in the field, wasn't necessarily enough to reduce inequities in her classroom. However, when Anne was supported with meaningful data and intentional strategies to address inequities that were observed, she was quickly and effectively able to shift the overall patterns of participation in her classroom.

3.3.4.2 Gwen

Gwen was an Asian American woman and tenured professor of mathematics education. As an educator, she is interested in student affect, as well as problem solving processes, creativity, and insight in mathematics understanding. Her teaching style was oriented around active student engagement, often through small groups and then whole class debriefs. She typically lets students group themselves, but also uses random grouping strategies to help students get to know other students. As an Asian American Woman, she was very aware of stereotypes, and for her, she sees the interconnections between race and gender as closely linked. Overall, her goal is to provide students with the support they need—which may differ between students—so that they can all reach the same end goal for understanding.

The first semester that Gwen participated, she was teaching a course on mathematics for social justice. Her course had 14 students, with 3 men and 11 women. The racial demographics were: 2 Black, 7 Latinx, and 5 Multiracial (all who were half-White). During her first observed lesson, there were a total of 41 contributions, and the gender breakdown was 17 men (41%) and 24 women (58%). Given that women made up almost 80% of the class demographics, this was a large skew in favor of men over participating. As Gwen became aware of the patterns, she was able to shift them with more intentional instructional moves. The shift in contributions can be seen in Figure 3.3.

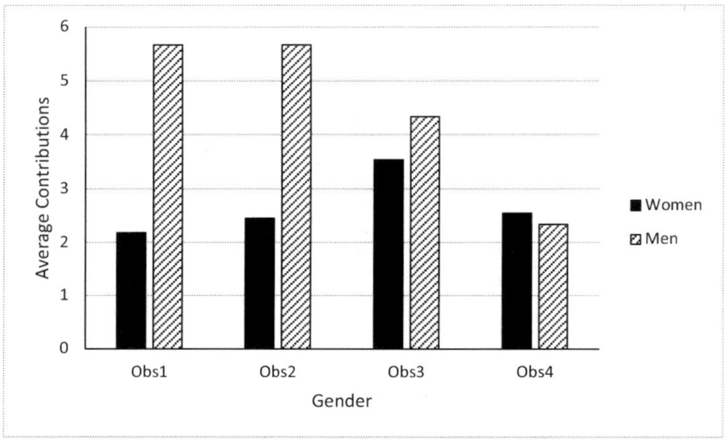

Figure 3.3. Average contributions by gender in Gwen's class over the course of the semester.

For Gwen, there were a few adjustments she made to her teaching that helped this shift. Typically, after students finished working in small groups, Gwen would call on groups to share their solutions. At this time, students within the group would spontaneously volunteer to participate, and it was almost always the same students. This led to two dominant men in the class taking up far more space in the classroom discussions than other students. After Gwen had data showing this, she began to use random methods to pre-select which student would share out from the group (e.g., the student whose birthday was coming up next). She also began to call on specific students with much more intentionality, which allowed her to shift this pattern. To support her to call on individual students without using cold calling techniques or causing stress, she tried to get to know her students better. For example, she constructed short activities that helped her get to know more about her students as people and used that as a foundation for deeper relationship building.

Gwen's shift in practice also had an impact on the racial dynamics in her classroom. In the initial observation, more than half of the contributions came from multiracial (mixed-White) students, who only made up 35% of the classroom demographics. Over time, Gwen was able to shift the participation patterns to be more equitable (see Figure 3.4).

3.3.5 Summary

Our work with EQUIP over the past eight years has allowed us to develop a methodology for tracking inequities in classroom participation. Across a wide array of settings, we have documented how racial and gender inequities permeate mathematics classrooms, and these studies are consisted with a large body of unpublished empirical observations. Most importantly, we find that through our professional development methods instructors can learn to shift such pat-

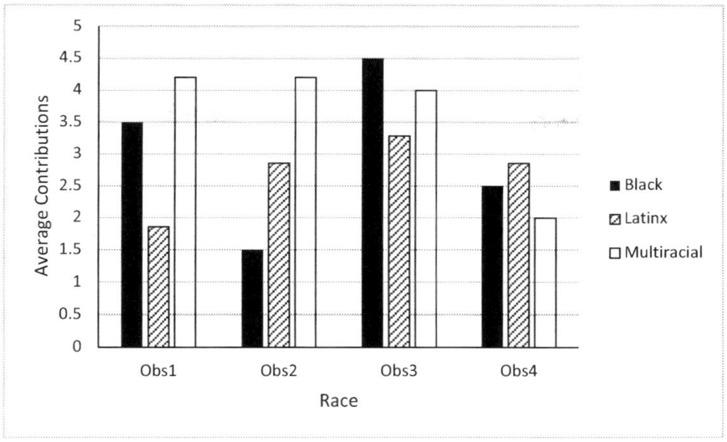

Figure 3.4. Average contributions by race in Gwen's class over the course of the semester.

terns. This book is designed to share many of the same instructional strategies that we use with participating faculty in a single, well-documented resource guide.

To be clear, these strategies are most effective when used in conjunction with data providing disaggregated patterns of classroom participation. While all the teaching strategies have merit and stand alone, they are even more effective when they can be used in targeted ways to disrupt inequities that are present. This allows for social marker specificity in the strategies that an instructor implements, so they can be intentionally used to disrupt hierarchies along axes such as race, gender, or disability. Consider the tools of Complex Instruction. Assigning competence allows an instructor to elevate the status of a specific student, but to make this most effective, instructors need a keen awareness of which students are perceived has having the lowest status, and that can sometimes be difficult without data making inequities visible and actionable. In Chapter 7, I provide guidance on how you could generate meaningful data to guide your own process of growth and instructional change.

3.4 Reflection

Now that you have reached the end of this chapter, here are some questions to support your further reflection. These questions will be most effective if you consider them in collaboration with others, who will have their own and different responses.

1. Think back to the guiding question of the chapter: in what ways is research about teaching mathematics to disrupt hierarchies consistent with what I already know, and what are areas for productive growth and learning to improve as an instructor?

2. Think about your own identities, backgrounds, and experiences. Reflect on how these have played out for you in different contexts. Have they mattered for you in some contexts more than others?

3. Think about yourself in the mathematics classroom. Which of your identities have mattered most for you in that space? How might your experiences be different if you had different identities?

4. Write a letter to a younger version of yourself. Now that you've learned more about this research, what is something you would tell yourself as a teacher to help when you were just getting started?

5. Looking at the references of this chapter, what is one reading that is new to you that you would like to read in its entirety?

4

Setting the Stage

The mathematics curriculum is bursting at the seams. It has been characterized as a "mile wide and inch deep" [281]. Many mathematics courses, like calculus, have a rigid, scripted curriculum that leaves little room for creative thought or deep engagement. As a result, many mathematics professors spend all their time on "covering content" and spend little time to support the social aspects of learning. Yet, building relationships, community, and creating collective access for your students is the foundation of disrupting hierarchies. It also requires changing the very content you teach, how students express their learning, and how you frame the mathematical enterprise. A common pitfall is to change everything at once, but my advice is to make small, incremental changes to your teaching, which will add up to a big difference over time.

As you transform your teaching, some of your practices may be unfamiliar to your students. For this reason, it can be advantageous to explicitly communicate to your students why you are teaching the way you are teaching. For example, you might share some research with your students about the science of learning [70, 89], and why your class will involve active engagement. Similarly, before building community standards, I might tell my students, "We are building community standards because we'll be working together a lot in this class, and it's important for us to agree as a community how we want to interact productively with each other." When we share our rationale for our choices, it allows our students to make sense of what we are doing, and, if we are open to it, it allows our students to give meaningful feedback to us.

Using education as a tool for liberation is an act of "radical love" [90]. Such radical love is a commitment to others, a hope for humanity, and the courage to act in ways to better the world. Other approaches such as abolitionist teaching [158] and disability justice [124] also emphasize the importance of joy, shared hope, and community care as a part of learning. These radical pedagogies make it clear that disrupting hierarchies and building a more just society starts with our own humanity and our relationships with others. If you want your students to function effectively as a community of learners, you will need to invest time, love, and energy into making that happen.

In the language of Discourses, for your students on the margins, they live in a world that devalues them every day [118]. Mathematics can be a profound site of dehumanization for students, dictated by cold, harsh, and competitive logics. In your classroom, you can intentionally create a different type of mathematical Discourse. Show your students that you value them, both in words and actions. This requires truly valuing your students for who they are, not coming from a place of saviorism or pity. Show up to your classroom with intention to transform the unjust world that we live in. In the sections that follow, we will consider how such a sentiment can be manifest in practice.

Before you read this chapter, I'd like you to take a minute to write down some of the strategies that you use to create a productive classroom environment. What are the conversations, tools, and practices that you use to build relationships with your students or create community? How much time do you spend doing these things? Please take out a piece of paper (or use an electronic device) and take five minutes to write these things down. This extra processing time will make it easier for you to see what is new to you. In addition, this creates an opportunity for you to write down *new ideas* in this chapter as you encounter them. As you read through this chapter, I want you to reflect on the following question: *what are ways that I can slow down my course to focus on the process of learning, and not just covering content?*

4.1 Relationships

Relationships are the foundation of teaching and learning. Learning requires practice, and practicing requires a willingness to try, experiment, fail, and try again, ad infinitum. Only when students are surrounded by people that they trust can they safely experiment without fear of failure. If students are afraid of judgment, they will be afraid to try, and it will be harder for them to learn. Unfortunately, many students have learned over years of experience that mathematics classrooms are not a safe place to experiment, and they fear being looked down upon by their instructors and peers [287].

A safe environment reduces stress. Research shows that stress impacts how we form, update, and retrieve memories, and in general, stress has a negative impact on learning [305]. When we are stressed, our bodyminds go into "fight or flight" mode, and our primary drive becomes survival, not learning. Stress also leads to a wide variety of mental health issues, so it should not be surprising that chronic stress can be a major barrier for our students. The negative impact of stress on mathematics performance is well documented, for example, when it manifests as a stereotype threat [15, 188, 277]. To reduce stress and build relationships with our students, we need to know them both as humans and as mathematicians. But first, we must show up authentically as ourselves.

4.1.1 Authenticity

Authenticity is foundational to relationship building and personal well-being [23, 30, 91]. Living authentically requires that one behaves and expresses emotions in a way that is congruent with one's beliefs, values, and identities. Showing up authentically to the classroom means showing up as a whole person, not a mathematical robot. When I was teaching in K-12 classrooms, I would sometimes run into students outside of class, in a grocery store or at the library. They would seem surprised to see me there, and their brains would have to register "oh, Daniel is a person too, not just a teacher." In the same way, we want our students to see us as people too, not just mathematics professors.

Here's a personal example of showing up authentically. Although I have been disabled all my life, I never allowed myself to bring that crucial part of my identity into my work life. I tried to pretend to be nondisabled, and hope that people never found out. But inevitably I would get very sick, or I would fail at masking my autism, and I would be greeted by surprise and a lack of empathy. In recent years I have been more open about my identities. In sharing my own experiences openly, I found that my students also responded with openness. Students were willing to share aspects of their identities—especially related to disability—that they would never have shared in the decade prior. My authenticity created the foundation for more authentic relationships with students. The openness of the students is notable, given that most disabled students have experienced discrimination in higher education. Students who are minoritized by other aspects of their identity (e.g., race) experience other forms of discrimination, all of which reduce trust and must be overcome to build productive relationships.

Authenticity requires vulnerability, which can also lead to us getting hurt. Overall, my experiences connecting with other disabled colleagues and building deeper relationships with my students have been positive. But to be clear, it has also opened me up more directly to ableist interactions. This can be difficult and disheartening. Interactions have ranged from ignorance to downright discrimination. While not all students will respond positively, overall, showing up authentically goes a long way to forming productive relationships with our students.

How we show up authentically also depends on our social marker identities. As a white and masculine presenting person, I'm afforded a high level of privilege in the mathematics classroom that makes it easier to bring in other identities without being questioned by students. Yet, many of my colleagues—especially women of color—share the many hurdles that they face with students questioning their mathematical authority because they may not conform to stereotypes that students have. Indeed, racial and gender biases that students hold against their instructors have been documented across a wide variety of settings [45, 83]. Consequently, these colleagues must put in additional labor—from the clothes they wear, the way they talk, their mannerisms, or their formality—to earn that authority from students, whereas it is given to others "by default" with much less work. Although authenticity is important for all of us, I want to be clear that some of us are afforded more opportunities (or easier opportunities) to show up authentically.

Authenticity is also not an excuse to discriminate against our students. I have heard colleagues use "free speech" or "academic freedom" as an excuse to engage in problematic racist and sexist behaviors in their classes. This is clearly not okay, and it will inhibit productive relationship building with students. At times one's personal beliefs may be at odds with our role as instructors, and at the end of the day, I come back to the adage of "not tolerating intolerance." Our classrooms should be an inviting place for everyone to learn, and if for some reason you believe that some students aren't capable of learning or don't belong in that space, I would challenge you to reconsider your beliefs. As instructors, we don't teach the students we wish we had, but we teach the ones who show up in our classrooms. I've heard many stories of instructors encouraging some students to quit math, and this can dissuade students who otherwise could have been very successful. It's easy for us to look back on our own experiences as mathematics students with rose colored glasses, but it can be harder to empathize with how our instructors may have seen us many years ago.

4.1.2 Knowing your students

Mathematics is commonly characterized as devoid of emotion. From this perspective, there is no reason to build relationships with students—we just stand and deliver the content. As a graduate student, I was told a story about a mathematics professor at my institution who once taught a course with a single student enrolled. Every day, the instructor prepared a formal lecture and presented it up at the board. The student never spoke. I don't even think that the professor knew the student's name. It should come as no surprise that I did not enroll in any courses taught by that professor. While this may be an extreme example, I have worked with many mathematics faculty members who have taught entire semesters and learned the names of only a few students. I doubt that those students felt seen by their instructor. Human connection requires authenticity, which cannot be achieved in anonymity.

Broadly speaking, when we build relationships with our students, get to know them, use their names, smile, and so forth, we create a psychologically safe environment. These types of interactions fall under the umbrella of *teacher immediacy*, which reduces the perceived distance between the instructor and students. A recent meta-analysis of studies of teacher immediacy shows that there is a modest, positive impact on student motivation [152]. Another meta-analysis found a significant correlation between both student affect and performance gains with nonverbal immediacy ($r = 0.49$ and $r = 0.17$, respectively) and verbal immediacy ($r = 0.49$ and $r = 0.06$), for small but measurable relationships [319]. Immediacy may also impact student persistence in college [319]. Moreover, some research in mathematics education suggests that the dominance of negative interactions—and lack of positive ones—is a factor contributing to racial inequities in urban schools [13, 12]. Clearly, how we relate to our students has a measurable impact on them and provides the basis for their learning. An first step to relationship building is learning the names of all your students.[1] This starts on the first day of class, or perhaps even earlier if you have a photo roster made available to you. Here are some strategies you can use to remember names,[2]

1. Create a photo roster of your students.
2. Have students create name cards/tents for their desks.
3. Have students say their name when they contribute to class.
4. Use student names frequently in discussions.
5. Talk to your students before and after class to practice names.

I once taught a course with 75 students, and I would challenge myself to greet every single student by name, every morning, as they came into class. A few weeks into the semester, I had learned all their names. Learning names doesn't come easily to me, but I can do it with practice. Using the above strategies will help make this process easier.

Early in the semester, I explicitly ask students to share their name each time they share an idea in class. "This is Maria, and my approach to this problem was to..." If you've ever been in a meeting with live captioning, this is a typical pattern of speech. By announcing the name of the speaker before sharing, it increases access for people who

[1] This may not be possible in a class with hundreds of students. Unfortunately, the teaching and learning process will suffer as a result, so even in those settings you should learn as many student names as possible.

[2] If you're in a virtual setting, you might even have student names right in front of you, but it's still helpful to practice remembering names as a part of getting to know your students.

may be visually impaired, or for students in the classroom who are seated somewhere that they can't easily see the speaker. This is a way to help students learn to see each other as valuable resources for mathematics learning.

Using student names strategically is also a very powerful strategy for broadening participation in your class. While it would be awkward to ask, "can we hear from the person in the red dress in the third row?" it is easier to ask, "Iris, I wonder what you think about this?" We typically learn the names of our vocal students without trying, but more importantly, we need to learn the names of our students who are less likely to participate spontaneously. Learning names helps us build relationships and intentionally bring those students into our class activities in a positive way.

Knowing a student's name is the first step to including them, and if you can't even learn the name of a student, how do you expect to know anything about their goals, interests, or mathematical thinking? It's also critical that we learn to say our students' names correctly (practice on your own if needed), and that we don't intentionally avoid using the names of certain students because they are less familiar to us. Don't reduce some of your students to nicknames! For people who have had their name repeatedly mispronounced, misused, or ignored all together, your lack of effort will be immediately evident. All of these are forms of microaggressions that diminish trust.

4.1.2.1 Knowing your Students as People

Learning names provides the foundation for learning more about your students. An easy way to get to know your students is to talk to them about things other than mathematics. You can do this before and after class sessions, or when circulating the room during work time. These casual conversations make a big difference. Get to know your students' goals, interests, challenges, and everyday experiences. This will also help you better relate to their lives as college students. Keep notes about different students if it will help you remember. The time you have before and after class, during office hours (which not all students will attend), and during breaks in mathematical engagement are all opportunities for human connection.

When we create a collaborative learning environment (e.g., with students working in groups), it creates space for us to connect with students in non-mathematical ways. While mathematical engagement is our primary focus, side-talk is a natural part of group interactions that we shouldn't ignore. The next time a student is mathematically disengaged (maybe even looking at their phone), rather than reprimanding that student (which furthers conflict), use it as an opportunity to talk to them briefly about something other than math. Then you can refocus, ask them where they are getting stuck on the math, and go from there. It helps build a more positive relationship than one of conflict and confrontation.

A mentor once told me that for each negative interaction we have with our students, we need three positive ones to balance it out. Recognize that many of our students have a negative association with mathematics. When we talk to them about mathematics, they may see that as a negative interaction. For such students, we need to find ample ways to build positive connections with them that aren't related to mathematics, so that they can trust us enough to take risks and move towards building a positive experience in our mathematics classroom.

4.1.2.2 Knowing your Students as Mathematicians

We also need to learn about our students as mathematics learners. This will help us build on their strengths and support them through their challenges. I start most of my courses by having students write their mathematical biography for me.[3] Here are three prompts I have used:

1. What is the story of your mathematical past? What are key events or experiences (positive or negative)? How do you feel about mathematics?

2. What are your goals for the course (not just passing the course)? What do you want to learn about and what do you hope to be able to do by the end of the semester?

3. What expertise will you bring to the course? It could be related to mathematics, or other skills that are useful to you as a mathematician. How can we create opportunities for you to use those skills?

[3] I was first introduced to this assignment by Brian Katz (BK) and want to acknowledge their influence on my practice.

Students can write a few paragraphs to answer these questions. They could also create a voice or video recording. This type of assignment is an easy opportunity for you to build multimodality into your course.

It's also valuable to share your mathematical biography with your students. You could do this informally, or you could formally write up your own biography. Personally, I like to share the story of my mathematical past, struggles, and goals for the semester with my students on the first day of class. I also share that I am a mathematics educator by trade, and as much as possible I aim to be transparent with pedagogical strategies I am using, including new things that I am trying this semester. Generally, I try to be as explicit as possible about the teaching strategies I use, so that students can understand the purpose of what we are doing.

When you engage with your students it is important that you show genuine interest in their thinking. Anytime a student says something—on topic or not—it provides a window into where your student is coming from. Have you ever experienced an instructor ask a question and then "fish" for a specific answer from the class? As a student I experienced this often, and it made me feel like the instructor didn't really care what we thought, they just wanted to fulfill their own mathematical agenda [172]. If you're going to ask questions to your students, then you should genuinely care what they have to say, correct or not. These student conceptions should form the basis of your instruction. When you interact with your students, structure the ways you interact with genuine questions, eye contact, and body language that communicates to your students that you care what they have to say.

4.1.3 Checking In

You should build regular check-ins into your course. These can be informal, and they need not be about mathematics. One way to do this would be to have a "check-in question" when students walk into the room. You could write something like "If you had to describe how you're feeling with an ice cream flavor, which would you choose?"[4] and have markers/chalk available for students to write up on the board as they come in. It doesn't take much time, and you can get a sense of how students are feeling. It can also build a sense of community. You can also add quick check-ins as a question on a homework assignment, or in the middle of group problem solving. These shouldn't take long, but they give you a temperature check on your students and communicate to your students that you care about how they are doing (which may be even more important).

I also highly recommend gathering formal feedback from your students at least once during the semester. Typically, I will create a short, anonymous feedback survey roughly 1/3 of the way through the semester. My favorite questions are these:

1. What is working well in this course?
2. What could be improved to better support you?
3. Is there anything else you'd like me to know?

Questions like this can give you invaluable insight. I try to do this early enough in the semester that I can still change my policies and practices, but not too early that students don't yet have a feel for the course. Doing this 1/3 into the semester (or about 4-5 weeks in) seems to work well. After getting responses, I share a summary of the feedback alongside my proposed revisions to the course and teaching methods. I ask students for feedback on the proposed changes. I also like to allow students to make their own suggestions for how things could be different. Although you typically may not be able to incorporate all feedback you receive, this is a valuable way to improve your teaching and increase reciprocity in your classroom. Incorporating a formal check in like this will communicate to your students that you care about what they think.

4.2 Community

In addition to the relationships that you build with your students, your students benefit from relationships with one another.[5] Fortunately, the effort you put into building individual relationships with your students provides a basis for

[4] I thank Katy Leigh-Osroosh for this check in question!

[5] At a meta level, we might also think about the need for us to have community as instructors, and the ways in which we learn from our peers. This is a core component of professional learning opportunities that I develop for instructors, which I touch on in Chapter 7.

helping students build relationships with one another. As you build community in your classroom, it will contribute to your students' sense of belonging, which is critical to persistence and success [148].[6] Student belonging has been studied extensively for students transitioning to college, and is significantly related to social adjustment and interest [120, 207, 297, 296].

You want your students to feel comfortable, connected, and excited when they come to your class. Your students will be much more successful if they are can build meaningful relationships with students who they can work with outside of class (e.g., in a study group).[7] Some students may build these connections independently, but for many, it is important that we intentionally structure ways for our students to get to know each other to build those connections that last outside of our classroom [119]. Research also shows that active environments can lead to more problematic interactions between students, which we will need to disrupt [59, 78, 266].

What we do in class communicates what we value. If we spend 100% of our time in class lecturing and working out problems for the students, we communicate that we think that they learn by listening to their professor. Similarly, what we test communicates what we value. If we assign myriad computational problems on homework, and then expect students to do conceptual work on an exam, we're mistaken – the homework already told them that we care about procedures and now we're sending mixed messages. It follows that if we want to develop a positive student community, yet we spend zero time fostering that community, we shouldn't be surprised when it does not manifest.

I've worked with many faculty who have the perception that students don't want to interact, that they want to just listen, and that they want to work alone. Many people interested in mathematics may identify as introverts. Our goal is not to force students to interact with peers all the time. Simultaneously, research is clear that interacting with peers in a cooperative setting is an effective strategy for improving learning [89, 272]. Moreover, many students will be working in collaborative environments in industry after graduation, so by intentionally structuring a positive collaborative environment for our students, we can model how to engage in productive working relationships. As I return to in Section 4.3, students may have a wide variety of different access needs, and we want to be mindful of prior trauma, social anxieties, or other differences our students might have. A guiding principle is to encourage all students to participate, but never to force students to participate. Some students may be really involved one day, but not on another, and that is okay. Simultaneously, we shouldn't use these individual differences as an excuse to uphold the racist and sexist participation patterns that are present in so many mathematics classrooms.

4.2.1 Community Agreements

In all of my experiences as a student learning mathematics, I never once had an instructor explicitly help us build community agreements for how we should interact. Either this wasn't deemed important, or it was assumed that students knew how to collaborate productively. This disadvantages your students who may not already know the rules of the game, or "hidden curriculum" to doing well in school [95]. For example, community agreements can make mathematical communication norms more explicit. Recognize that there is a cultural and linguistic aspect to how people interact, and you may have students coming from many cultural backgrounds where the standard norms of engagement are different. Moreover, students have been socialized through prior experiences, and may have certain expectations about what a mathematics classroom is that are inconsistent with the type of environment you aim to create.[8]

On multiple occasions I have seen instructors provide a set of "threats" to their class: NO PHONES, NO TALKING IN CLASS, COPYING HOMEWORK WILL RESULT IN FAILURE. These are not community agreements. Not only does threatening your students not work (students who want to cheat will find a way to do it), it creates a hostile environment that removes safety and inhibits trust. Suspicion Discourses of this nature are disproportionately

[6] How you approach this depends on students' prior relationships. If students already know each other as a part of a cohort program, you will want to think about how to leverage those existing relationships.

[7] Research shows that having a supportive community to collaborate with to solve mathematics problems outside of class is a huge factor for success. One key example of this is the Emerging Scholars Program (ESP), which was developed at UC Berkeley [92]. The ESP builds community between students with additional collaborative problem-solving sessions. These aren't framed as remediation, but as an honors program. In particular, the program was designed to improve Black student success at Berkeley. The impacts of the program were profound, reducing failure rates from 30-40% to less than 10% [92]. This model has been replicated successfully at a variety of other institutions [119].

[8] Thanks to Ernesto Calleros for sharing his in-progress dissertation work that helps delineate these linguistic, cultural, and social aspects of communication in undergraduate mathematics.

targeted at marginalized groups of students, for example, when disabled students are frequently considered to use accommodations not for access but as a way to cheat [191].

I use the term community agreements instead of "community standards" or "norms," to emphasizes that students have an active role in authoring these agreements. Setting up an initial conversation around community agreements should happen very early, ideally the first or second day of class. Typically, I provide students with an interesting mathematical problem to work on collaboratively for 15–20 minutes. After that experience, I facilitate a reflective group discussion:

1. What went well?
2. What could be improved?
3. What are the implications for our work together as a class?
4. What helps you succeed as a learner?
5. Can we come up with a set of standards or agreements on how we should engage with each other?

Although I always let students lead this process, I reserve the right to discuss certain suggestions or make my own suggestions. Another way to organize this process is to first give students 5–10 minutes in small groups to discuss agreements (sans the collaborative problem solving), and then discuss as a whole class. This discussion should take a minimum of 10–20 minutes to allow for depth of thinking. Here's a sample of community agreements from a course I recently taught on undergraduate mathematics education:

- Listen to others and be respectful of their opinions/thoughts
  - Give your attention to whoever is speaking
  - Critique ideas but not people
- Be mindful of dominant voices
  - If you talk a lot, leave room for other people to talk more
  - Be aware of your own identities and how they fit into dominant narratives
- Recognize people are at different places in their growth journeys related to some topics
  - Everyone learns at a different pace. Be patient, mindful, and kind.
- Have multiple access points to math content (technology, pen paper, etc.)
  - Make your thinking accessible/easier for others in your group to see.
- Understand that mistakes are a part of the learning process.
  - Encourage critical thinking but don't belittle ideas
  - Don't be a hater
  - No shame in sharing what you're thinking
- Encourage team members to contribute, ask each other questions.
  - Make sure you're using your power to bring others into the conversation
  - Make sure that everybody contributes to the conversation and make sure that their ideas are taken into consideration.
- Revoice each other's ideas to ensure understanding.
  - Restate what has been said, because you have a slightly different perspective
- Take care of your body and what it needs
- People need think time (sometimes individual)
- Create structured ways to share and check in

This list is intended to provide an example, but not necessarily a model to replicate. The agreements your class comes up with should be responsive to your local context. Before having this conversation, you'll want to create a positive, open, and safe environment that gives students the space and comfort to share their thinking.

4.2.2 Revisiting Agreements

Once established, it helps to regularly revisit the agreements. One way to do this is to create a poster or PowerPoint slide that can be projected to students, for example, before groupwork time. This allows you to remind students of community agreements that were established for how they should work together. You can strengthen this approach by following up the group work time with a short (approximately five minute) discussions about how things went. In what ways did they follow the standards? And what still needs work? By taking time out of your class session to discuss process, you're communicating to your students that process matters. In an online setting, you could so something similar with a shared document and debrief the observations you make in breakout rooms. When I teach, I like to keep a clipboard on hand so I can take notes about what students are doing in my classroom. For a situation like this, I craft up a quick chart that has three columns: (+), (−), and "Quotes" [116]. I jot down examples of productive collaboration (+), things that could be improved (−), and any quotes I want to share ("Quotes"). When highlighting positive engagement, I always use students' names if I know them, and when talking about negatives, I don't use names to avoid calling anybody out. Even if you do this only 2-3 times in a semester, it really brings home the point that process matters to you, and you wish to help students engage productively.

4.2.3 Community Fun

One of the best ways to build community is for your classroom to be fun. Fun isn't a distraction from doing serious mathematics, but rather, it should be a part of the process. If you frequently use group work, let your groups name themselves to build shared bonds. Provide your students with opportunities to move around and engage in embodied mathematics [276]. Allow for laughter, jokes, and enjoyment. The more that students find mathematics as something that they can enjoy, the more likely they will be to persist with it.

4.3 Access

A positive classroom community is one that meets the access needs of all members of that community. Access means that everyone can participate in the space in their full humanity and have their varied needs respected and met. An accessible environment will afford disabled people with opportunities to acquire the same information, engage in meaningful interactions, and utilize the same types of services that a nondisabled person can in an equally effective way. Disabled people should be able to learn and obtain information equally and independently as someone who is not disabled.

Ensuring access is both a matter of accessible design (i.e., flexible practices that generally work for a lot of different people), and specific accommodations for individual needs that are not adequately addressed by your overall design. When others attend to our access needs, it communicates care and mutual respect. In contrast, when our access needs are ignored, it inhibits trust and safety. Disability justice reminds us that access matters for everyone, not just disabled students [124]. Access needs are many and can be varied, including:

1. Physical access and ergonomics
2. Opportunities to move, stretch, and take breaks
3. Food, drinks, appropriate bathrooms
4. Language/communication that is comprehensible
5. Safety, trust, and nonjudgment

This is a short list, intended to demonstrate that access needs cross a variety of identities. Consider access to food, drink, or bathrooms. A student may have a medical condition that requires them to be able to eat frequently. Or a student could be a breastfeeding mother. A student could be fasting for religious reasons (e.g., Ramadan). A student may need access to gender-neutral bathrooms. Regardless of the reason, creating a flexible space that can meet these access needs shows respect and care for students.

Communication is another important area of access. While American Sign Language (ASL) and captioning are common examples that come up, communication access needs are varied. For emergent multilingual students, the use of idioms or overly complicated sentence structures could create barriers [176]. By making intentional design decisions–such as providing language supports through consistent talk routines, technology, or written and visual supports alongside verbal directions–we can improve language access [326] Similarly, for some disabled students, formats like Easy Read or Plain Language can communicate ideas more clearly [43].[9] In general, writing down instructions and providing as many ways as possible for students to understand the tasks (e.g., by helping them understand a problem context), is a positive alternative to only explaining a task verbally without support.

A broad conception of access requires problematizing the elitist and exclusive nature of mathematics. Mathematical proof is often taught as a rigid set of conventions with only one correct way to communicate. This is clearly false. Proofs are not merely ways to establishing mathematical facts, but they are ways of communicating mathematical ideas [103, 283]. From this perspective, our proofs should aim to help others learn mathematical ideas, rather than confuse them. While pictures are often not included in a formal proof, visual representations alongside formal logic can enhance conceptual understanding [235].

It is important that we don't use "mathematical convention" as an excuse for exclusionary, ableist, racist, and sexist practices. Within the realm of mathematics, certain types of cultural knowledge (which don't conform to Western rationality) are deemed invalid and ignored. For women and racially minoritized students, creating an environment that is welcoming, supportive, and free of microaggressions (or macroaggressions) is a baseline for access. Here I outline a few strategies for creating access. Disrupting microaggressions is an issue I return to in Section 5.7.1.

4.3.1 Syllabus

For many instructors, the syllabus is the only place that they formally discuss access with their students. Unfortunately, most universities take a legalistic approach to access. This begins with a boilerplate statement in the syllabus about how disabled students must request accommodations from the student disability center, and the professor is only responsible for meeting those accommodations with formal approval from the disability center. This practice is problematic for many reasons. Before the semester even begins, it pits disabled students in opposition to their instructor. Especially given negative stereotyping and stigma associated with disability, students must weigh the costs of coming out as disabled. Moreover, many students have experienced having their accommodations denied by professors. For example, in a survey of 114 students who received accommodations at a university, 35% of them reported that one of their professors outright refused to provide them with legally mandated accommodations as dictated by the Americans with Disabilities Act (ADA) [175].

Disabled students also need to expend additional energy in self-advocacy to receive accommodations, rather than focusing on just learning the material [129, 211]. When students do self-advocate, they are often viewed as lazy or trying to cheat the system. The situation is even worse for students who do not have access to a formal diagnosis. Receiving a diagnosis is often expensive and time consuming, and that is for people who can even access a provider who can provide a diagnosis. In other cases, students might be entitled to accommodations and not realize it.

As a point of comparison, I personally spent ten years going to hospitals across the US to receive a diagnosis for my autoimmune condition, and I was in my 30s when I learned I was autistic! I didn't receive a single accommodation from kindergarten to completing my PhD, but in retrospect, I definitely could've benefited from some. There are also many students, like me, whose disabilities don't fit into the limited menu of cookie-cutter accommodations available. Often, students themselves might not even know what possible forms of support could help them be more successful. In mathematics, accommodations are often limited to testing settings, and this can lead to problematic power dynamics [190]. At the end of the day, these syllabus statements inform students about their legal rights as mandated by the ADA, but they may do little to create access within your classroom.

Rather than including a boilerplate statement in your syllabus doing nothing more, I recommend including two statements. As is required by law, you should include an "Accommodations Statement" that informs students of their legal rights under the ADA. I also recommend including an "Access Statement,"[10] which describes how you aim to create access in your classroom. Rather than hiding this near the end of the syllabus, you could put this up front to highlight your commitment to access. Without such a statement, the solitary Accommodations Statement will mostly communicate to your disabled students that you do not care about them. Here is some possible language you could

[9] Here's an example. Easy read uses short sentences. Simple words are used when possible. Easy read avoids jargon.

[10] Sam Ridgway would also call this a "commitment to access."

consider for an access statement, but you should modify it according to your needs and commitment to creating access:

> My goal is to create a space that provides access to all students in my classroom. I recognize that all humans have access needs, and I am committed to helping you meet those needs in my classroom, whether you are disabled or nondisabled. You should be able to participate in this space in a way that fully respects your humanity, allowing you to act in ways that are necessary to care for your body without fear of violating an unspoken social rule. I aim to create flexible assignments that will provide a variety of different ways for you to engage, learn, and demonstrate your understanding. While I strive to make this space accessible to all, I recognize that you as an individual know best what you need for access and how you learn. I invite you to share any particular access needs that you have to help you do your best work, and you do not need to disclose (or have) a formal diagnosis for me to respect your access needs.

Albeit imperfect, the statement communicates a general understanding of different access needs individuals may have and a willingness to work with students to support them. Of course, if you put something this in your syllabus, you also need to enact it with your practices.

4.3.2 Accessible Design

Universal Design for Learning (UDL) emphasizes that students benefit from multiple means of engagement and multiple ways to show their learning [247]. Rather than relying upon a narrow set of learning activities and then providing students with accommodations, UDL provides guidance on how to create activities that create access by design. UDL is all about choices. It involves providing multiple means for engagement, representation, and action/expression.

A large body of research focuses on the value of multiple representations in mathematics [61]. In short, having more different ways to represent a concept (e.g., equation, table, graph, verbal description) provides multiple means for developing a robust mathematical conception. This can help improve student understanding [26]. Simultaneously, multiple representations provide different entry and access points for different students and are therefore a useful part of accessible design.

Another practice for accessible design is multimodality. Mathematics education research (including my own) has focused extensively on verbal participation, but there are many ways to participate. When students are involved in verbal arguments, creating written inscriptions, drawing pictures, enacting concepts through embodied movement, gesturing, using manipulatives, and so forth, it provides them with different means to engage with the concepts and build more robust understanding. In fact, the use of multiple modalities is suggested as a best practice by NCTM [204]. Although mathematics prizes abstraction, having multiple ways to ground abstract concepts makes them much more concrete and accessible to a variety of people. Even for professional mathematicians, the understanding of abstract concepts tends to be supported by a wealth of visualizations and other types of examples.

4.3.3 Movement and Physical Access

Research shows that humans have limited attention spans and can only listen/focus for a limited amount of time. This may be especially true for students who have ADHD or who are otherwise neurodivergent. This has implications for accessible instructional design; we should break up our class sessions to allow a variety of forms of access. Often, this will have students getting up, moving around, moving desks, etc. In situations like this, we also want to be mindful of physical access in our classroom space. For instance, are people stuck in rows where they can't get up to use a bathroom? If you have a wheelchair user, can they navigate the space? Are students uncomfortably close, or too far away, from their peers when working together? While certain aspects of the physical layout may be out of our control, to the extent that we can, we should try to make a physically accessible classroom for our students. The more we know about our students' access needs (even shared privately), can help us do this more effectively. Other alternatives to having students move around include pausing for a short "wake up break" (even just 30 seconds), and allowing for students to fidget without stigma.[11]

[11]For students with ADHD or autism (and perhaps other students too), fidgeting (with or without the use of a fidget toy) can actually help them focus better. This is a form of *stimming* or self-stimulation, in which the use of some sort of repetitive movement can be used as a way of blocking out external distractions and actually focusing better. As an autistic person and a drummer, I find that I often use drumming or tapping on surfaces as a way to better focus, and I do this unconsciously without realizing it. Sometimes teachers put too much emphasis on sitting still, in place, with eyes focused on the board, without recognizing that other forms of paying attention exist and may work better for some of your students.

4.3.4 Flexible Assessments

The standard mathematics course typically consists of high-stakes exams and homework sets. However, if competence is defined this narrowly (i.e., completing procedures quickly and accurately), it creates barriers to access. It is also inauthentic [314]. Mathematics research is messy, involving peer review, revision, and ongoing reflection [113]. By choosing more open-ended tasks, we can provide a variety of ways for our students to engage (see Section 4.4.1). As a guiding principle, consider how you can offer voice and choice to your students. Flexible assessments are all about giving options to empower your students. If your goal is for students to understand a particular concept, can you think of different ways that your students could show that understanding? You can also engage your students as active participants in this process. Allow them to suggest possible alternative ways to display their competence. Chapter 6 is all about assessment, so I offer a wealth of concrete suggestions there.

4.3.5 Accommodations

Although we should strive to create access for all our students by default, some of our students may require specific accommodations in addition to the environment we have strived to make accessible by design. Common examples involve ASL interpretation or alternative text or captions for images. Accommodations may constitute anything that a student needs to access learning. Given that society is ableist and receiving a diagnosis is a privilege that not all students have, my policy is to provide reasonable accommodations to any student who asks for them, regardless of a formal diagnosis or letter from a student disability center (e.g., a student may be dealing with sickness, or may be dealing with a family hardship impacting their mental health). Access needs can relate to various identities, including culture, language, religion, etc. And when we build our learning environments without careful thought, we ourselves create new access barriers for our students (e.g., lecturing for 50 minutes straight doesn't support focused attention, or putting students to work in groups without proper support can result in problematic microaggressions). While students don't always know the accommodations that would help them the most (when I was a student, I didn't), I take as a guiding principle that students know a lot more about themselves as learners than we do as their instructors, who are just getting to know them. For this reason, it's important to talk to our students, ask them about what they need to have access, and be willing to periodically revisit those conversations as student needs (or awareness of those needs) changes over time.

4.3.6 Normalizing Access Talk

Providing one's pronouns when introducing oneself has now become a common practice. In disability justice circles, it is also common to share one's access needs [237]. This may feel awkward at first but is something that can be learned and normalized. For example, I might say "my name is Daniel, I use they/them pronouns, and my access needs require me to be able to move around, take frequent breaks, and ensure that I am not exposed to illness, because I am immunocompromised." Stating access needs does not require disclosure of a medical condition or life circumstance, it just is a matter of sharing what is needed to "do one's best work." Students should never feel forced to disclose something they are uncomfortable disclosing or don't want to share with peers. You can practice sharing access needs as students introduce themselves. For students who are having all their needs met, they can say "my access needs are currently being met." It is not appropriate to state that one does not have access needs, because we all do. It is simply that some of us are privileged and used to having our needs met on a regular basis, while others are not. This may be easy to do verbally in a small class (less than 20 students), or you may consider asking about access needs by surveying students in a larger class.

Access check ins are also an important tool for creating access. Access needs are fluid and can change from day-to-day or moment-to-moment. Practice taking breaks in class and asking questions like the following:

- Are access needs being met?
- Do folks have access to this video I'm playing?
- What do you need to do your best work?

If you make time to ask questions like these, be prepared that your current plans may not actually provide access to your students. You should aim to be flexible and change your course of action as needed. In addition to these in-class check-ins, you can allow for anonymous feedback on class surveys or through exit tickets, to ensure students have plenty of opportunities to share their access needs. You can include language such as "I care about your needs as a student" or "it is important for me to create an environment in which you feel comfortable." By directly signaling that you care about your students and their needs, you can help them feel more comfortable to share with you honestly.

4.4 The Math

The final aspect of setting the stage is the mathematics we teach. One way to conceptualize mathematics is through a list of topics to be covered in a content domain (e.g., which chapters of the calculus book does your class engage with?). This is a narrow conception of mathematics content, and it's not very productive for helping us improve our teaching. I believe that most mathematics courses would be improved considerably if roughly half of the topics were thrown out, and the other half were engaged with more deeply.[12]

Narrow conceptions of mathematics content are often paired with narrow conceptions of understanding, as an answer either being "right or wrong" for a standardized assessment. These conceptions focus little on processes, and don't reflect authentic mathematics. When we move beyond covering content, we can attend to the types of mathematical practices that our students engage with. We should also think deeply about how students learn a particular content area. Finally, how we frame the mathematical enterprise (i.e., our epistemology of mathematics) has a huge impact on how our students perceive the discipline and their relationship to it. Given that mathematics is a highly stereotyped discipline, we can work to explicitly disrupt those stereotypes.

4.4.1 Mathematical Tasks

Although it is not the primary focus of this book, I want to talk briefly about mathematics tasks. If we want students to collaborate and engage with peers in our classroom, we need to choose tasks that are capable of fostering collaboration. It helps when we have tasks that are open, complex, and allow for many different forms of engagement. Overall, these features describe problems that will allow for productive collaboration, discussion, and are more likely to elicit a variety of forms of student thinking. For example, in an open ended task or a mathematical modeling task, students would first need to discuss the assumptions that they made in approaching the task.[13] These effective problems have been described as *group worthy* tasks [155]. Group worthy tasks are messy and complex enough that they require the skills of multiple students working collaboratively to be tackled effectively. Given this messiness, such tasks create space for the different competencies of a variety of students to emerge, setting the stage for status interventions like assigning competence (see Section 2.3.7). Drawing from this concept and adding a few ideas from my own experience, I offer the following features of tasks that are effective at promoting collaboration and engagement,

- They are open ended and require real problem solving (not just following a procedure).
- They allow for multiple solution paths (or multiple solutions).
- They connect to big disciplinary ideas like making conjectures, arguing, proving, problem solving, modeling, and conceptual understanding.
- They provide opportunities for students to integrate concepts or representations.
- They prompt students to explain and justify their ideas.
- They allow students to contribute by drawing on a diverse set of skills and strengths.
- They have a low floor (accessible to all students) and high ceiling (allow for students to go deep in their thinking).
- They are fun, engaging, have real-world implications, and connect to students' lives.

[12] Whether our institutional and peer constraints actually allow us to do this, is another question altogether.

[13] By contrast, if we ask our students to work out procedural computations, it rarely supports productive collaboration because there isn't very much to talk about. Students typically do the procedure correctly, or not.

Realistically, the tasks you choose won't meet all of these criteria. However, when you can choose tasks that meet more of these criteria, you will find that they support more productive collaboration. Choosing such tasks that promote collaboration is also one way of pushing back on ideas that mathematics is individualistic and competitive.

Equity is also a consideration in our choice of tasks. For example, there is a large body of literature documenting how some standardized assessments have in-built forms of racial [88] and gender [57] biases, as well as culturally-based language demands [1]. Instructors often assume that all students will understand the context of a problem and the language used in the problem, but that's not necessarily true. When we use real-world contexts that are more familiar to some students than others, we privilege them by allowing to draw on their own cultural knowledge more effectively. This doesn't mean that we should only use problems devoid of context (because context can help students connect mathematics to their lives), but rather, we should take care in ensuring that students understand the contexts. Some design principles for achieving this include: 1) aligning the context across multiple problems (e.g., using the idea of running races as a recurrent theme to talk about position, velocity, and acceleration); 2) including language goals in addition to content goals; and 3) using common language routine, talk moves, and structures that can become more familiar to learners over time [326].

Even mathematical tasks that appear equitable on the surface may rely upon cultural biases (e.g., what does a "usual" apartment look like?) which privilege certain types of cultural knowledge and experiences over others [194]. By framing mathematical tasks more openly (e.g., as not having only one correct answer), you can create space for various cultural knowledges to be valued. For example, the *multiple-ability treatment* from Complex Instruction (see Section 2.3.7) involves describing a variety of skills that might be required to effectively do a task, which can help a greater proportion of students see themselves as having something meaningful to contribute to group work [50]. In contrast, if you frame tasks in terms of getting the single right answer, it is more likely that cultural biases could be reinforced, and fewer students may see themselves as having meaningful opportunities to contribute.

The way that we use a task also has important implications for how it is taken up by students. We can understand this idea through the concept of *cognitive demand*, which focuses the types of mathematical thinking that students get to do [282]. Tasks with lower cognitive demand focus on factual recall and executing procedures. Tasks with higher cognitive demand focus on making connections, providing justifications, and offering explanations. Some research shows how the implementation of high cognitive demand tasks can promote student interest [189] and learning [112]. This is broadly consistent with more in-depth research on self-explanation and testing effects, which have been widely studied [73].

Another way to think about cognitive demand is who gets to do the interesting mathematical thinking. Is it the instructor, or the students? In proving a major theorem, do you do most of the work for the students and leave them with trivial computations, or do the students do the heavy lifting? How we frame a task can change its cognitive demand. Even in graduate-level mathematics courses, I observe that instructors often reduce cognitive demand and leave students providing low level responses. This reinforces the idea that mathematics is about getting the right answer, and it also eliminates many opportunities for interesting mathematical thinking that your students could have engaged in.

Suppose that we begin with a mathematically interesting task, like the *four fours* problem. This problem asks students to find mathematical expressions for other numbers using a combination of four copies of the digit four and other mathematical operators. For instance, $4/4–4/4 = 0$ would be a valid way to express zero. An open statement of this task might ask students "Which numbers can you express with four fours? Can you express all numbers?" Such a statement of the task does not reveal which numbers can be expressed, as it is part of the exploration. In contrast, a more closed framing would say "Find expressions for all numbers from 0 to 100 using combinations of four fours." Another closed framing of the problem would be "It is possible to express any number using a combination of four fours and appropriate mathematical operators (hint: how could you use logarithms to help you?)" In using these more closed framings we can lessen the cognitive demand, because students get to do less of the mathematical thinking.[14] Thus, when we think about choosing tasks, it's also important to think about how we enact those tasks, to ensure that students get to do the meaningful thinking.

[14] In some cases you might use a more closed framing if you wanted students to practice a particular procedural skill, for example.

4.4.2 Mathematical Practices

Regardless of the tasks you use, you should aim to portray mathematics authentically. Doing real mathematics is messy. Hersh [113] contrasts front and back door mathematics. Think about dining at a fine restaurant. When you enter through the front door, you are greeted by a refined environment, and you are served delicious, well-prepared food. However, if you were to enter through the back door (i.e., as a chef), you would be entering a much more chaotic environment. Food is being prepared, people are shouting across the kitchen, and you are doing what you can to take raw ingredients and make something delicious. All this chaos is needed to support the refined experience through the front door.

Mathematics is much the same. The "front" of mathematics is a well-organized set of theorems and proofs that are neatly arranged to present mathematical theory. This is what most students encounter in school. But much of the real excitement happens with the "back" of mathematics. This type of mathematics is messy, informal, intuitive, and full of failures and missteps. It's full of exploration, conjecture, and drawing pictures. If students are to ever have any hope of producing the neatly packaged mathematics that we see in the front, they are going to have to spend a lot of time getting messy working through mathematics in the back.

Mathematics in the back is the place where mathematicians engage in authentic practices. It doesn't say a lot about you as a mathematician to know that you can memorize the strategy for every related rates problem in a calculus textbook. Solving a problem when you already know the answer isn't really problem solving. It's just factual recall. If we want students to experience us as mathematicians, then we need to put ourselves in situations where we may not always know the answer. This doesn't mean that we should come to our classes unprepared, but rather, we should be prepared to model an authentic inquiry process for our students. It can be very valuable for us to let our students see how we struggle through a problem, get stuck, and get unstuck, because these are all normal parts of the work. No matter how much mathematics you know now, once upon a time, some of the very same mathematics you are teaching was hard for you. Try to remember what it was like when the problems were hard and get in your students' shoes.

This is a first step towards creating an atmosphere where it is okay for students to explore, make mistakes, share false starts, and get feedback from the instructor and their peers. When we create unstructured, open-ended learning opportunities, our students can develop more productive beliefs about what it means to do mathematics [257], and they can develop the metacognitive and reflective skills are needed to effectively navigate these messy situations [99, 256]. Give your students plenty of opportunities to practice and get feedback from you (and their peers) by intentionally structuring this messiness into your classroom. Mathematics is beautiful. Doing mathematics is a form of art. Share the art with your students [153].

Broadly speaking, the messiness of doing authentic mathematics has been framed in terms of mathematical practices [203] or mathematical habits of mind [64]. These practices include things such as making conjectures, exploring open-ended problems, proving and argumentation, modeling, and generalization. These are the authentic types of things that a mathematician or someone working in a math-intensive industry does. When I talk to mathematics instructors, these are the interesting types of things that they want their students to be doing. However, there is a disconnect, because often these very same instructors are teaching in a procedural way that focuses on covering content and does not afford students opportunities to develop these practices.

4.4.3 Normalizing Failure and Productive Struggle

Research shows that social engagement is crucial to learning. As students explain their ideas, receive feedback from peers, and become a part of a community of learners, it will deepen their understanding of mathematics. Because participating is so closely connected to learning, the students who participate the most are the ones who benefit the most from the learning opportunities available. Crucially, students don't need to be correct to learn from their participation. In fact, sharing in-progress ideas is the best way to get meaningful feedback that can support learning. Making mistakes is a part of learning. We only learn when we are challenged, and if we aren't making any mistakes, then we clearly aren't challenged. Students should prepare to fail early and fail often.[15] The road to success is paved with failure.

It is easy to say that we value failure, but it is more challenging to implement this in practice. One strategy you can use is called "my favorite no" [36]. This strategy involves having students solve a problem on an index card, which is

[15] I have to attribute this statement to my mentor Alan Schoenfeld, who repeated it often!

turned in to the instructor. The instructor quickly sorts the cards into correct and incorrect piles and tries to find one card that has a common or interesting mistake. This solution is discussed with the class, both focusing on what was done well in the solution, and where the mistake was made. Such a strategy honors the importance of partial attempts in developing understanding. Another way to do this is to have students create a "failure portfolio," in which they collect mistakes they have made, and have an opportunity to think through how they have overcome them. I talk more about portfolios in Section 6.4.1.

Students learning mathematics are like athletes training for an event. For an athlete to run farther or become stronger, they need an appropriate challenge. They need a training plan that pushes them to run just past their limit. They need to lift heavier weights. If they try to push too hard, they might get injured, but if they don't push hard enough, they won't make any progress. Progress comes through the progressive overload of muscles. The body adapts in response to an appropriate amount of physical stress on the muscles. The job of a coach is to design a training program that provides just that level of challenge. Without struggle, there is literally no progress. An athlete knows that if their regiment is too easy, it has little value.

Learning mathematics is like training the mind. No growth will happen without an appropriate challenge. In mathematics education, this is often framed as "productive struggle," when students struggle to make sense of mathematics and understand something that is not immediately obvious [310]. Productive struggle is also related to an optimal range for learning that has been called a "zone of proximal development" [306], in which a learner engages in a social situation with appropriate support, pushing their thinking beyond what they could do alone. There is a common misconception in society that an "easy" learning process is a sign of intelligence, but this is false. Regardless of where students are in their learning, it should be appropriately difficult for them. As an instructor, you can help your students better understand how learning works, and frame the challenges not as deficits, but a natural, necessary, and inevitable part of learning. To be able to do this effectively, you need to know a lot about what their students and what they can do, so you can push them just past the limits of their current understandings.

4.4.4 Mathematical Knowledge for Teaching

Beyond knowledge of the content itself, instructors also benefit from knowledge of how students learn that content. More formally, this type of knowledge is called Mathematical Knowledge for Teaching [115]. Suppose you're teaching a calculus class. At a baseline, you need to know how limits work, how to solve problems with them, and perhaps have fluency with formal proof. But to teach effectively, you'll need more than that. You will also benefit greatly from some understanding of the ways in which students typically think about limits, which aspects of limits may be conceptually challenging for students, what resources students bring to the table, and how to create effective activities that support deeper thinking about limits. These types of understandings are unnecessary for being a successful mathematician, but they are crucial for being an effective mathematics instructor.

Developing Mathematical Knowledge for Teaching takes time. Especially if you're teaching a course for the first time, or if you're teaching a course that you only teach infrequently, you may not be very familiar with how students think about that content in general. If you're lucky to be teaching with high quality curricular materials,[16] the materials may include common types of student thinking or sample work. This type of curricular support—and even tasks that were explicitly designed with this knowledge of student thinking—can greatly enhance your ability to elicit student ideas and productively build on them. Most textbooks are not infused with knowledge of student learning in this way, so you may think carefully about the materials available to you and how you build on them.

Regardless of the course you're teaching or the textbook you're using, you can build opportunities to understand student thinking through your teaching. The ideas of eliciting and building on student thinking are the focus of Chapter 5. Ways to elicit student thinking through assessment are the focus of Chapter 6. As you make student thinking a focus of your course—perhaps even by having students explicitly analyze the thinking of peers—it provides you with a pathway to developing knowledge of how students learn a given content area, which will greatly enhance your teaching of that content.

[16] An exemplar that comes to mind is the formative assessment lessons from the Shell Centre (https://www.map.mathshell.org/lessons.php). In higher education, the TIMES Project has created materials for Differential Equations (https://iode.wordpress.ncsu.edu/), Linear Algebra (https://iola.math.vt.edu/), and Abstract Algebra (https://taafu.org/ioaa/). These are just a few examples and the list does not aim to be exhaustive.

4.4.5 Epistemologies of Mathematics

Mathematics is a human endeavor. But few students experience it that way. Most students go through school experiencing mathematics as a disconnected sets of facts and procedures that just exist. They don't see it as something that humans create (or discover, depending on your epistemology of mathematics). And most certainly, they don't see it as something that they could create themselves. Personally, I take a social constructivist view of mathematics as a cultural endeavor and something created by humans [79].[17] As we think about broadening mathematical engagement to create space for more people across different intersecting identities, it stands to reason that even more different types of mathematics will be created.

If you want to model mathematics as a human endeavor, first and foremost, you should model yourself as a human who is on their own mathematical journey. When my students first see me, they see a mathematical expert who they assume has loved and has always been good at mathematics. They don't see me copying from my neighbor to pass eighth grade algebra, dropping out of high school, or struggling and barely making it through introductory calculus. They don't know that I didn't fall in love with mathematics until my senior year of college. I share details like these, at the appropriate time, so my students can see me as a human who also does mathematics. If we can't empathize with all the ways that mathematics can be hard for our students, then it will be difficult to help them work through and overcome their struggles.

It's also important to humanize the subject matter that we're teaching. We can do this by providing historical context to the discoveries and people who made them. We do this by making connections to real-life issues (e.g., gerrymandering, global climate change) and mathematics. We can do this by providing opportunities to create their own conjectures, explore them, and try to create their own mathematics. Mathematics is so important to our lives, yet we often teach it in a way that is devoid of context. Even in "pure" mathematics subjects, there are humans involved in every step of the process and we can teach about them. There is history. There is a reason why we use limits in calculus, even though it may seem counterintuitive and mostly unnecessary for the type of work we're doing in the classroom. Help your students understand the why. Teaching the history of mathematics development—and humanizing the people involved—also provides an opportunity to discuss the messiness and politics of who is celebrated in history, and whose contributions are erased.

4.4.5.1 Humanizing through Role Models

It is important for mathematics learners to have roles models that are like them. As a disabled person, I did not know a single disabled mathematician or mathematics teacher that could be a role model to me, so I simply assumed that the only way I could be a mathematician was to try to be nondisabled. This is impossible, and it put two important identities of mine in conflict, and this was a problem. Students from any minoritized identity group may feel similar conflicts [164].

Depending on your own identities, you may be a role model for some of your students. If you show up authentically to your classroom, the students who resonate with that authenticity will feel a stronger sense of connection. For students with other identities, you can provide other types of role models. For example, you may have programs within your institution that support minoritized mathematicians and scientists, and you could invite students from those programs visit your classroom to share work with your mathematics students. This can be the start of a longer-term relationship to disrupting hierarchies in your institutional space.

Famous mathematicians can also serve as role models. Related to the process of humanizing mathematics, it can be powerful to bring in the stories of historical mathematicians who are relevant to your course. For example, a colleague of mine incorporated the biography of Sofya Kovalevskaya into one of his courses. There are myriad interesting mathematicians – Emmy Noether, Maryam Mirzakhani, Evariste Galois, or John Urschel, to name a few – who defy expectations about who a mathematician is or what they do. These are fascinating stories for students to learn, which unfortunately, they almost never get to learn.

[17] I also recognize that throughout history, many mathematicians have (and even today) take a view of mathematics as a perfect body of knowledge that exists in a higher realm of perfection (i.e., a Platonist account). I do not believe that accounts of the historical development of mathematics across cultures support this view, and moreover, I think this viewpoint can also reinforce the idea of mathematics as an exclusionary endeavor.

There are also numerous communities that have sprung up to celebrate the mathematical contributions of contemporary mathematicians from different communities. These include: Mathematically Gifted & Black, Lathisms, Indigenous Mathematicians, Spectra, and Sines of Disability, to name a few. The free books *Living Proof* [111] and *Testimonios* [105] also have a wonderful collection of powerful stories of modern mathematicians. Social media are also treasure troves full of stories that show there are so many ways to be a mathematician. Bringing these into your courses with intentionality can help break up the problematic, stale, racist, and sexist stereotypes that exist.

4.4.5.2 Deconstructing Stereotypes

Stereotypes are generalizations about groups of people, that are often grounded in false cultural narratives, expectations, and mass media, which may be driven by less-than-benevolent motives. These stereotypes implicitly affect the ways in which we act, even if we don't expressly believe in them. This relates to the concept of implicit biases. Even though we might believe all students are equally capable of learning mathematics, if we only see a certain type of student succeeding in mathematics, implicitly, we make the association that those students are better at math. This then impacts the ways in which we interact with our students and becomes a self-fulfilling prophecy.

For me personally, I share with my students that I am a mathematics education researcher, that I do work around Discourses and classroom participation, and that we know certain stereotypes are ubiquitous in society, and especially mathematics. These include ideas that mathematical intelligence is innate, some people simply aren't math people, or that certain groups of people (like Asians) are the best at mathematics. Your students all know these stereotypes, so that there is no point in pretending that they don't. If we acknowledge those elephants in the room, then we can work to address them. Once we surface these ideas, we can talk about all the ways that they are false.

One of my favorite ways to surface stereotypes about mathematicians is to have my students complete the *draw a mathematician* task [213]. To introduce the task, I ask my students to take out a blank sheet of paper and spend five minutes drawing whatever comes to mind when they think about a mathematician. This is an excellent way to see how your students are thinking about broader stereotypes about mathematics. When I have my students do this, about half of them draw stereotypical images, like an old disheveled white man with glasses. They describe this person as frustrated, sleep-deprived, and lonely. The other half draws something different. Some students might draw themselves. Others draw what they want a mathematician to look like. Regardless of what your students draw, this is an excellent way to talk about stereotypes in mathematics. If you can project students' images while they talk about them, it is easier to show all the different viewpoints to your class.

This task is also an excellent exercise for showing genuine interest in student thinking. You might use questions such as the following to frame a discussion:

1. What did you draw?
2. Why did you choose that?
3. Do you agree with that stereotype?
4. Why do you think those stereotypes exist?
5. What would it take to change those stereotypes?

It's important to give students a chance to talk about how they are thinking about their drawings. You can use this as a lead in to talking about why stereotypes may exist, why they are problematic, and how you are creating a different type of environment in your classroom.

4.5 Reflection

Now that you have reached the end of this chapter, here are some questions and activitities to support your further reflection. These questions will be most effective if you consider them in collaboration with others, who will have their own and different responses.

1. Think back to the guiding question of the chapter: What are ways that I can slow down my course to focus on the process of learning, and not just covering content?

2. Think about a space that you feel comfortable and welcome in. What are the qualities of that space? How is it similar to or different from your classroom?
3. Imagine you're planning a community gathering (e.g., a dinner party). People from a whole variety of identities are going to be there. How would you plan this event to be inclusive? How would you think about people's access needs? What other considerations would be important for you? How does this thought experiment relate to your classroom?
4. What are your access needs and can you think of an example where somebody did not respect those? How did you respond and how did it feel?
5. What is an example of productive struggle in your own life? What did you learn from that process and how did you persevere? What support did you have, and what support did you wish you had?
6. Over the next week, try to do something you've never done before in your real life (take a safe risk). What did you need to prepare to do it? How did you feel after you did it? What would make you want to try it again? How can this help you relate to your students who are learning something new?
7. What was the most interesting math problem you ever worked on? What made it interesting?
8. Find a child in your life and talk to them about what makes a mathematician.
9. Gather as many exciting art materials as you can. If you had to see yourself as an amazing mathematician, what would it look like? After completing your drawing, try to deconstruct any identity-based stereotypes you see in the art.
10. Do a syllabus review with your students (do this sometime in the middle of the semester). Have a discussion with your students about what you're doing, the perspectives that you've been bringing, and what you're learning so far. Have a reflective discussion with your students and talk about what could come next.
11. Commit to talking to one of your students at a more personal level and learn something about their life. What did you learn? How was that experience?
12. Try to use at least three student names during a class session. During the next lesson, use three different names. What did you notice?
13. Looking at the references of this chapter, what is one reading that is new to you that you would like to read in its entirety?

5

Facilitating Practice

In Chapter 4, I focused on setting the stage to support student learning. In this chapter, I focus on how you organize opportunities for your students to participate in meaningful mathematical practices. These strategies will allow you to control the flow of classroom participation and to confer meaningful learning opportunities upon students who may face multiple barriers to participation. Although you want to create rich opportunities for all learners, you will need to offer special attention and use strategies to support your students at the margins, especially those who are multiply marginalized (e.g., across race, gender, disability).

A key assumption of this chapter is that you know something about the distribution of student participation in your class, based on student social marker identities. While all of these strategies can be used in their own right, unless you use them with intentionality (i.e., with social marker specificity, by attending specifically to student identities), they won't necessarily make your classroom more equitable. If you wish to disrupt hierarchies in your classroom, then you need to have awareness of what the hierarchies are. While there is commonality across settings and classrooms, how power and privilege play out in any given classroom will have its own unique features. Hence, the more data you can gather to support your teaching to disrupt hierarchies, the more effective you will be. How you can generate such effective data to guide your learning process is the focus Chapter 7.

Overall, when these teaching strategies are used in conjunction with social marker specificity to acknowledge and attend to the different experiences minoritized students have in our classrooms, it allows us to draw out the ideas of our minoritized students and publicly position them as competent. This process consists of three steps: 1) identifying a student perceived as low status, 2) identifying their mathematical contribution, and 3) making that contribution public.[1] The data you collect about student participation will allow you to identify a student perceived as low status. To identify mathematical contributions, you will need to use strategies that elicit student thinking (both whole-class and small group or partner formats), and you will need to target those strategies towards minoritized students. Finally, you will need to make that competence public in front of the class, to raise the status and support further engagement from specific minoritized students (thus reversing patterns of historical inequities that are present in your classroom, as evidenced by your local data).

Before you read this chapter, I'd like you to take a minute to write down some of the strategies that you use to facilitate conversations. Do students talk in your class? How much? What strategies do you use to get them to talk? How are you able to manage who talks, and what concrete practices do you use that attend to student identities in the conversation? Please take out a piece of paper (or use an electronic device) and take five minutes to write these things down. This extra processing time will make it easier for you to see what is new to you. In addition, this creates an opportunity for you to write down *new ideas* in this chapter as you encounter them. As you read through this chapter, I want you to reflect on the following question: *What is one practice that you would like to commit to trying the next time you are in the classroom teaching students?*

[1]This three-step process of assigning competence was described in this way to me by Charles Wilkes II.

5.1 Structures

The structure of our class sessions provides the foundation for participation. Different forms of participation support learning in different ways, and crucially, they provide opportunities for different students to participate. I've observed countless classrooms in which whole-class discussion is the only format that is used to structure student participation, and this constrains the opportunities available for students to learn. Here, I provide a general overview of different ways you can structure participation, and the benefits and drawbacks of each. The four basic participation formats involve students working: as individuals, with partners, in groups, or as a whole class. Productive engagement in any format is supported by clear guidance on how to participate. For example, if you want students to work with peers, you could say something like:

> Okay, now we're going to share our idea with a peer. I'd like you to turn your desk to the person next to you and have a conversation for one minute.

With a statement like this (and the physical act of moving a desk), you're communicating a clear expectation (with support) for students to engage. If some students don't follow your directions, you can check in with them individually. You might say, "Hi Stephen, I noticed that you didn't turn your desk. Is there a reason for that?" If Stephen responds that he didn't know who to work with, you could help find an appropriate partner. If Stephen responds that he wants to work alone, you might respond, "In this class you will also have plenty of opportunities to work alone, but I'd like for you to try to chat with a partner right now. Sharing your ideas out loud and getting feedback can give you a new way to think about this content." If Stephen is still insistent on working alone, I wouldn't force him to work with a peer. We don't always know what is happening in our students' lives, and something major may have happened that puts Stephen in a place where he's simply not ready to participate (e.g., a death of a family member). Usually, I find that with a little encouragement most students will choose to engage in partner work. You can also ask "What would make this learning opportunity more accessible to you?" to identify potential barriers to address.

I often observe that when instructors are new to supporting student collaboration, they don't provide clear directions on how to participate. They might say something like "you can talk to a peer if you'd like." This doesn't tell students whether you think it is important for them to talk to a peer. It also doesn't provide access to students who might be hesitant to reach out to a peer to talk. This puts the onus on students (especially marginalized ones) to participate in a partnership that may not feel safe to them, and there are no clear parameters to help make it a safer or more productive conversation. Try to avoid statements like this. Instead, provide clear directions for students to choose a partner, or use a strategy to choose the partners for your students. If you are choosing to structure participation in a specific format, it should be because that format is well suited to the learning goals you have for that activity. Explaining the goals of that activity and why that structure is a useful one can go a long way to helping students see the value in what you're doing.

5.1.1 Individual

Mathematics lectures often move at a fast pace to cram as much material as possible into a small amount of time. This provides students with little opportunity to pause, reflect, and process the material that you are teaching. As a mathematics graduate student, I recall the sinking feeling that I wasn't keeping up with the pace of a lecture, and the knot in my stomach as I contemplated asking my professor to pause or slow down. Questions ran through my mind. *Was it worth taking the risk and possibly irritating the professor or peers in the classroom who had more mathematical experience than I? Would it reflect poorly upon me as a mathematics learner?* In retrospect, I probably wasn't alone in my feelings that I couldn't keep up with the lecture, and I'm sure that many of my peers could have benefited from a pause as well.

Pausing is the simplest way to incorporate individual think time into your classes. Intentionally schedule pauses and short breaks for your students to process material, and for you to plan your next move. Rather than asking "do you have any questions?" try telling your students,

> I'm going to pause for a minute so that you have time to process what we just talked about. I'd love for you to share and questions or reflections that come up for you when we come back together.

Research shows that incorporating intentional pauses into your instruction will significantly increase the proportion of questions that students can answer and will increase the length and depth of those responses [300]. For more detail look at Section 5.4.4. Pausing is also a great way to allow processing time for students who are learning English or may have auditory processing issues.

In general, implementing individual work time during your class sessions is useful if there is something that you want all students to do, and it won't be that beneficial for them to engage with peers yet. For example, if I wanted students to perform a computation or work a single step of a problem, I would choose for them to do it alone, because this helps build their individual procedural fluency. By circulating around the room, I can diagnose which students might be struggling with that specific procedure. Moreover, if you've ever listened to students discuss procedural problems, it's not particularly interesting. One student is usually telling the other how to do it, and there's little justification why to do it that way. Or you might hear two students arguing about their different procedures, but they have no way to resolve the conflict. Overall, there's just not much for people to productively discuss in pairs or small groups about disconnected procedures, so creating such a structure to discuss procedures is not a great use of time and resources.[2]

Individual time can also be used productively in conjunction with other modalities. Suppose you want students to discuss a concept in pairs, but you want to make sure each student has had a chance to first think about the idea before talking to a peer. This is the basis for the routine "think-pair-share," which as the name suggests, structures a conversation by starting with individual think time, talking in pairs, and then sharing as a whole class. A rationale for having students think first before sharing is that it helps avoid a situation in which some students might just wait to hear an answer from a peer, rather than trying to figure it out themselves first.

5.1.2 Partners

Partner work is a special case of group work, in which all groups are size two (except perhaps a single "pairing" of three if you have an odd number of students in your classroom). Partner work allows students to engage with peers without the same level of complexity of social interactions that occur in larger groups (i.e., group dynamics). Through social interactions, students get to explain their thinking and they can get feedback on their thinking. As discussed earlier, this can tap into many of the learning benefits of "tutoring effects" which have been shown to have a significant impact on learning across a wide variety of settings [82, 132, 246].

Moreover, when each student is only working with only a single peer, students typically have more opportunities to talk, because they are sharing the conversation with only one student rather than many students. Of course, sometimes a single student dominates a peer conversation, but overall, partner work allows for a genuine exchange between students and all students can engage more frequently than in larger groups. In my own research, I have found that partner discussions provide meaningful opportunities for all students to talk [233]. Nonetheless, I did identify small but significant differences in the proportion of talk, by race and gender, with non-minoritized students speaking more [233]. Similarly, research on pair programming in computer science—a context in which partner interactions have been comparatively well studied—shows that partner interactions provide a relatively balanced set of opportunities to participate, but subtle inequities do emerge along the social marker dimensions that one would expect [147, 269]. This contrasts group work situations, where some students are more likely to be completely marginalized [49, 138]. This is a common occurrence in my professional development research, in which we routinely observe groups in which one or two students are barely involved in the conversation.

Using explicit framing can help prevent a single student from dominating a partner discussion. For example, you might say,

> The first partner will talk for a minute to share their idea, and then when the timer goes off, the peer will have a minute to respond. Please don't interrupt your peer during their talk time.

There are numerous ways that you can structure the conversation. More important than any particular structure is that by giving explicit instructions you can help ensure all students get to talk.

[2]Although it can sometimes be helpful to practice procedures in class, in general, I would argue that working on disconnected procedures is not a particularly effective use of class time, because you are not leveraging the social aspects of learning that can be enhanced by having many learners working together in a community. Procedural practice is better left to independent practice, and I would most likely discuss procedures in class briefly with the goal to diagnose and remediate possible issues students are having.

If I want students to engage with peers, and I only have a small amount of time, I default to partner work due to its flexibility and ease of implementation. For example, if I am presenting a proof, I might have students turn to a peer and justify why a step works, clarify a definition we are using, or try to come up with an example that fits with our general argument. These types of quick checks build a sense of community between the students and help surface whether everyone is following along. If students have questions, it is less threatening to ask them to the whole class, because they can ask a question on behalf of their group, rather than as an individual. Moreover, because the flow of discussion has already paused, the student doesn't have to take the risk of "interrupting" me. This is can help create access for students who might have anxiety around participating or who are afraid of fulfilling negative stereotypes.

Partner work also works well for moderately complex problem solving. In these situations, the one-on-one atmosphere makes it much easier for students to ask clarifying and probing questions to one another, which again they may not feel as safe doing in a larger group setting. With students working in pairs, I can easily move around the room and observe the myriad different types of thinking, before bringing the class back together. It's also an easy way to check in with specific students who I know I want to check in with. Once I have broken the class into partners, I can intentionally focus my time on students that I know I want to attend to. This allows you to provide extra support to some students without publicly drawing unwanted attention to them.

I also like to use partner work to allow students to provide structured peer feedback (more on this in Section 6.3.2). In this peer feedback setting, students can easily swap partial solutions with each other, read them over, and then exchange feedback. Having conducted my own research on peer feedback for many years, I have found that sharing feedback in this way doesn't really work so well in larger groups, because too many students are sitting on the side just waiting for their turn.

Lastly, partner discussions also allow for a check in before a whole-class discussion. This could happen as a part of the think-pair-share routine described above, or skipping the individual think time, and just having a "turn and talk" with a peer before the whole-class discussion. A short partner discussion before a whole-class discussion helps ensure that every single student has at least something that they can share to the whole class. By asking students "what they talked about" rather than "what the answer is" any student can be brought into the public sphere. When you regularly use these types of conversational shifts—focusing on process rather than getting the right answer—you communicate to students that you value process over product.

As should be clear, partner work is flexible and relatively easy to implement. It has great learning benefits, and it helps sidestep some of the inequities that arise in small groups and whole class. For this reason, I strongly recommend the regular use of partner work, especially for instructors who are just beginning to get their students more actively engaged in their class sessions. Especially if a course uses similar pairings frequently, an instructor can also add in processing time to reflect upon what is working well and what might need to be reconfigured within students' working relationships. Such reflections can lead into a whole-class discussion that revisits community agreements and whether and how they are being enacted in peer engagements. Not all pairings will be equally effective, and if you can get feedback from your students, that may help you avoid pairings that don't work as well.

5.1.3 Groups

When done well, group work with unstructured problems can be a useful tool to help students develop skills for authentic mathematical problem solving [256]. This helps students see the messiness of mathematics in the making [113]. Group work is appropriate when you would like (ideally 3–5) students to work on messy, open-ended problems (see Section 4.4.1) for an extended amount of time (15–30 minutes, or more). Although group work can be very beneficial for students, it takes much more setup to use effectively compared to partner work. For this reason, if a problem is simple enough that it can be handled in individual or partner work, I strongly recommend you use one of those modalities. Group tasks are often complex, challenging, and time consuming, so it helps to foster a notion of productive struggle (see Section 4.4.3).

Sometimes during group work time you may notice that a lot of student groups have the same question. You can address this using a *catch and release* strategy. Rather than continuing to answer this question to individual groups, the second time it came up you could use a mechanism to get the attention of all students (e.g., talking loudly, flicking the lights off then on again) and answer that question for the whole class. This will save you time and will help make groups work better. In general, if you find students are struggling or getting off-task, a quick catch and release can bring the class together, refocus, and then help them explore more in their groups. In this way, you may find that sometimes you will move between group work and whole-class in a way that supports group engagement.

Group work consists of free-flowing interactions between students, and as a result, is a site in which power and privilege manifest [81]. Group work is also a space in which toxic interpersonal interactions can arise [59, 78, 266]. Microaggressions are a form of identity-based threats, like the suggestion a woman can't do a particular proof because "men are more logical than women." Although these may not manifest directly in front of us as instructors, research shows that these problematic interactions are too common in mathematics classrooms. Even if you don't witness such interactions directly, you can assume that they may be present in your classroom, and therefore, you should take extra care to address issues when they do arise. By intentionally structuring group interactions and community agreements you can help mitigate these issues before they happen. Later, we will discuss strategies for addressing these types of issues when they do arise.

5.1.3.1 Group Composition

Although every classroom is unique, in general, the more axes along which a student is privileged, the more likely they are to dominate small group conversations [49]. This can be understood by considering the ways that systems of oppressions are mutually constitutive and interlocking [56]. For this reason, I have observed in my research—across hundreds of mathematics classrooms—how men in general, and white men specifically, tend to dominate the participation in small groups. This has to do with both racial and gender socialization, in which white masculinity socializes students to be louder, aggressive, and more competitive, speaking over their peers. These contrasts other ways of knowing that are more collaborative and inclusive. Empirically, research documents that as the number of men in a group increases, the masculine presence (and thereby marginalization of women), also increases [66, 266]. As a result, women have significantly more opportunities to participate when they are in a group of majority women, rather than a gender-balanced group or one in which women are the minority [66, 266]. Moreover, in unstructured collaborative group settings, gender roles emerge spontaneously, relegating women to more traditionally feminine roles (e.g., recording the work). For example, a variety of studies have shown how men take up disproportionate opportunities to use technology such as Augmented Reality (AR) [309], laboratory equipment [216], or a computer [266]. Although these studies take place in STEM fields other than mathematics, it's not much of a stretch to imagine how they could take place in a mathematics classroom (e.g., a computational computer lab).

Knowing this, it is important to be mindful of gender composition anytime your students will be working in groups. In general, putting a single woman in a group with three men is probably not the best idea. Research on the impact of other identities in small group interactions is less clear, but in our research, we have documented how racial inequities arise in favor of White and some groups of Asian students [266]. While I cannot provide clear guidance on group composition based on race, language, disability, and so forth, these identities matter in small groups. Watch out for groupings that could surround a single minoritized student with students from dominant groups, because there is a high likelihood that inequities will arise.

So how do we arrive at ideal configurations for small groups? If we leave students to choose their own groups, some students may struggle to find a supportive group and feel left out and marginalized. Students might also seek out groups based on friendships, but these don't always lead to the most productive mathematical interactions. Similarly, random assignment may result in problematic arrangements.

In my own teaching, I implement two less-than-ideal solutions. First, I often use pseudo-random assignment. That is, I will randomize the names of students in the classroom, but then if I get non-ideal arrangements (a single minoritized student with three white men, for example), I will reconfigure the groups. Depending on the demographics of your class—especially if there are only a few minoritized students—you may consider placing them in the same group (at least sometimes), rather than splitting them up so they are all outnumbered in other groups. I have heard many instructors say that they would explicitly split those students into separate groups "to expose them to other ideas," but I believe this is a mistake and it may lead to those students feeling singled out (or divided and conquered).

Another strategy is to give voice to the students. Even if students do not directly choose their groups, they can still have input into their groups. One way to implement this is randomize groups throughout the first few weeks of class so that students are working with a wide variety of peers. Then, a few weeks into the semester you can survey your students to provide input on students they are working well with, and others who they would prefer not to work with. This can give you some guidance for creating groups in your class.

There are other variables to consider. Should your groups always stay the same, or should they change? A benefit of consistent groups is that the students can get to know each other better and build positive interdependence. Also, in my experience, this almost certainly leads to at least one group that is not very functional, which gets behind, and doesn't learn as much. For this reason, I try to keep groups semi-stable (for at least a few weeks at a time) and mix them up infrequently throughout the semester. This allows some benefits of stable social relations but also breaks up those groups that get stuck in a rut. How exactly you manage these group configurations is something you can experiment with to see what works best in your local context. Creating a short questionnaire or Google Form that asks students questions such as:

1. What's working well in your current group?
2. What could be improved?
3. Would you like to change groups?
4. Are there any students you would prefer not to work with?
5. What else would you like to share?

5.1.3.2 Group Roles

In addition to forming groups, we want our students to work productively in their groups. Explicit conversations around community agreements and checking in with students about process are some effective strategies to promote this. Another strategy is to have group roles (e.g., note-taker, whole-group reporter, and equity monitor). These roles feature heavily in instructional strategies such as Complex Instruction (see Section 2.3.7) [50]. Roles serve dual purposes of giving every group member an important way to contribute to the task, as well as elevating productive behaviors for collaborative learning. Roles help delegate authority from the teacher to students in an explicit way, which allows them to engage in more substantive classroom interactions. Especially for students (or an instructor) who are less familiar with group work, this can be a helpful entry point.

Personally, I don't have as much experience with using Group Roles in college math classes, but I have a few colleagues who have been very successful. For example, my colleague Chris Rasmussen uses them in his teaching. I've even spoken with his students who shared what a positive experience it was. Chris used the following set of roles:

1. *Starter.* This student is responsible for getting whatever materials are needed and helping the group get focused on the task. This student also puts away any materials at the end of group work.
2. *Moderator.* This student ensures that everyone in the group gets to participate, and when students share an idea, that everyone listens and understands that idea. This student attends to group equity.
3. *Reporter.* The reporter is responsible for synthesizing and sharing the group's work with the rest of the class during a whole class discussion.
4. *Spy.* The spy has a special role. During designated times, all of the spies move around the room to check in with other groups to get ideas about what they are doing.

Each class session, students in the group are assigned to take on different roles. This ensures that all students have at least some opportunities to take on high-status roles (such as reporting out to the class), as one way of disrupting classroom hierarchies, and also promoting more productive collaboration. Students in the class work collaboratively on difficult tasks using shared white boards. Chris has the students write their roles on the top of the whiteboards, which also allows him to bring some students into the larger whole-class conversation with intentionality (e.g., he could explicitly bring in a student who is assigned the role of reporter by asking that group to share their ideas).

Another colleague of mine, Johanna Rämö has also been using group roles for many years in Finland. She shared a current iteration of her group roles with me for inclusion here. She has her students track their use of different roles throughout the semester, and all students must take on each role at least once.

1. *Scribe.* The scribe is responsible for writing up the group's work, or for asking other students for assistance in writing.
2. *Reporter.* Reports to the teacher and to the other students in the end discussion what the group has done and thought about.
3. *Information seeker.* Searches for the information needed to solve the task in the course material and online.
4. *Critic.* The critic brings up unclear issues and presses the group to more deeply explain their reasoning and provide justification.
5. *Scheduler.* Makes sure that the group moves forward with the tasks at a suitable pace (doesn't get stuck on one thing for too long, but on the other hand, doesn't rush too fast).

While the exact roles that you use may vary (and can be a part of your own experimentation), the fact that you have structured roles and that you are teaching students to use them helps support the collaborative process.

Even if you do not formally assign roles to students, you might observe when students continue to fall into the same roles (i.e., the same student always has the marker, the same student always shares out group strategies) and explicitly prompt students to mix up their engagement (e.g., "Marcus, why don't you give another student a chance to work on the white board?", "When we come back to whole-class, I'd like to have someone share out who didn't share out last time.") If you do plan to use student roles, it is something you should do regularly, and plan for time to help students learn how to fulfill these roles [50].

5.1.4 Whole Class Discussion

Whole-class discussions are ideal for synthesizing mathematical ideas because an instructor can draw out ideas from all members of the class, rather than just a subset of them. This can allow for a productive compare-and-contrast of ideas, which allows for students to synthesize different perspectives and deepen their own understandings. It is also more efficient than attempting to summarize the same material multiple times across different small groups of students.

Simultaneously, participation in whole-class discussions tends to be inequitable. I've empirically documented this across hundreds of classrooms [78, 238, 228] and in my professional development work [236, 231]. It is challenging for even the most skilled instructors to facilitate an equitable whole-class discussion. Doing so requires practice and explicit strategies to mitigate inequities (by race, gender, and other identities). I spend a lot of time mathematics classrooms, and rarely do I observe an equitable whole-class discussion.

One way to promote more productive and equitable discussions is to provide students with meaningful opportunities to do some mathematical thinking prior to the discussion, for example through partner or group work. During student work time, you can check in with different students about their thinking, and then invite them to share their thinking later when you come back as a whole class. This allows you to intentionally choose student ideas to connect across concepts and representations. Having multiple representations and multiple ways of thinking about the same topic leads to deeper conceptualizations [72]. Moreover, when these discussions are held in relation to specific learning criteria, it helps students build stronger skills of evaluative judgement [292]. Research-based strategies also support bringing in student ideas from small groups into whole-class discussion, which has a significant impact on their participation and status [49].

If you invite your whole class to participate in a discussion, you want to ensure that every student could offer something to that conversation. If you start with a whole-class discussion when you first introduce a new topic, some students will have familiarity with the topic, and others will not. Thus, using a whole-class discussion at that stage will further privilege the students who have already had more exposure to the mathematics in your course. Although you may feel satisfied after having a compelling conversation with the one or two students who already have familiarity with the topic, most of your class will be left as spectators who don't really follow along. In my research I have frequently observed one-to-two students (often white men) take up most of the talk time, not leaving space for other students. You can avoid this by using intentional facilitation strategies, which are discussed throughout this chapter.

Whole-class discussion are also a powerful venue for reshaping status hierarchies in your classroom. While it matters for students to be seen as competent in small group or partner settings, the stakes are much higher in a public, whole class setting. Many students many fear contributing to the whole-class discussion because they are afraid of being humiliated publicly. Being publicly seen as competent is important for most humans, especially in mathematics, which is used as a tool of measuring smartness and sorting people. As an instructor dedicated to disrupting hierarchy, you can flip this around and use whole-class discussion to reshape status hierarchies (e.g., see Section 5.5).

5.1.5 Additional Participation Structures

The above structures – individual, partner, small group, and whole class – are the four primary participation structures that I use on a regular basis. Each format has a different purpose for learning, so it is helpful to use all of them. Sometimes, you might be looking for something different to shake up the flow of your classroom. Here are a few additional structures you might try.

5.1.5.1 Gallery Walk

A gallery walk is basically what it sounds like; you create a gallery of student work and allow everyone in the class to walk around and observe the gallery. This only works if you have sufficient wall space and your room allows for movement. In practice, I usually have students work in small groups on an open-ended problem or discussion prompt, and then have them create a visible artifact to share with their peers. This could be work written up on blackboards on the wall (if you're teaching in such a room), or it could be written up on large sheets of poster paper. Once the groups have created their posters, you give some amount of time (about 5-15 minutes) for everyone to move around the room and see the work of others (being mindful of physical access, see Section 4.3.3). I like to provide my students with guiding questions, such as:

1. What do you notice in how other groups approached the problem?
2. What do you wonder about how other students approached the problem?
3. Do your answers agree? Can you mathematically connect the differences?

These questions are just a starting point. Depending on the mathematics your students were working on, you would want to customize your questions accordingly.

The gallery walk is useful because it gets students moving around which can help spur their thinking. It also gets students away from sitting passively in their desks and builds an environment in which students know they will be active and engaged. Research shows that the longer students are sitting and passively listening in their desks, the more their attention and focus begins to wane [32]. In a gallery walk, you can overhear many interesting conversations as students walk around the room and interact with a variety of peers. As you circulate around the room, you have targeted opportunities to talk with a variety of students. One variation on the gallery walk is to have a "poster session." This would mirror a poster session at a professional conference. You can have one member from each group stay with their poster to discuss with others, while the rest of the group members walk around. Then you can switch it up so different people stay with the posters, and everyone has a chance to circulate.

5.1.5.2 Everyone Shares

Another strategy for including everyone in the classroom is to set up an "everyone shares" situation. This involves every student sharing something very briefly with the class, generally on the order of a single sentence. This often can be paired effectively with another participation format as a means of synthesis. For example, you might set this up by saying:

> I would like everyone to share one thing that you noticed during the gallery walk. Everyone in the class will have a chance to share, and I want you to keep your observations to a single sentence.

Quickly moving around the room this way will give everyone an opportunity to speak and is a low stakes form of participation. You could imagine asking students to share something for a test review, a struggle on homework, a key takeaway from a lesson, and so forth. There are many ways to use this as a quick and inclusive mechanism for synthesis and taking the temperature of your classroom.

A variation on this theme is the six-word memoir/story. A six-word story tries to communicate a lot, in just six words. "Six words can say a lot." This is a fun way to have students consolidate thinking in a different way. For example, you might ask students to write a six-word story that summarizes a key idea from the day's lesson. You could also have students write a six-word story describing how they are feeling about how things are going in the class. The opportunities are endless, and this is a fun mechanism to check in with.

5.1.5.3 Paper Toss

A paper toss essentially involves people throwing lots of paper around the room. This is a useful technique for sharing lots of student ideas around the room. Sounds fun, right? To set up for the paper toss, you have students answer a guiding question, and write down their answer. This could be about a mathematical argument, a question they have, or a key idea from class. Students write down their answer or question on a loose sheet of paper that they crumple up and toss around the room.

In the next round, each student picks up a piece of paper and reads it. Now you have options. You could simply have the students crumple the paper and throw it again. Or you could give students an opportunity to add to the paper. For example, you could ask them to write a question that they have in response to what was written. Now, students crumple up again, rinse, and repeat. You get to determine the number of tosses when you use this method.

The paper toss introduces an element of randomness and anonymity into your classroom (in a low-tech way). Students don't write their names on the paper, so nobody knows who has written what. It allows students to see the thinking (or questions) of their peers, which can build a better sense of community between the students, or camaraderie if the students are feeling stuck around certain mathematical ideas.

After your students have tossed the papers around, you can use this as the entry point to a discussion. You might ask students to share one idea that they read, or what connections they made between papers. You could ask students to share one question that you saw someone else write down. The possibilities are numerous, because students get to report on the thinking of others, and suddenly the fear of failure is taken out of the equation. You can pull off this whole paper toss event in about five minutes.

5.1.5.4 Whiteboards (or Blackboards)

My favorite rooms to teach mathematics in are the ones that have whiteboards (or blackboards) everywhere. I want students to write on the walls, tables, or any other surface that allows for it. This helps make work public which supports collaboration and sharing out.

You can also create your own whiteboards [234]. Whiteboards can be expensive, but one cheap way to procure many of them is to purchase melamine panels (like those used to back cabinets) and have them cut to the appropriate size. Most hardware stores will do the cutting for you. These panels will make customizable dry erase boards for only a few dollars each. Students can engage in both scratch work and group work on this, so this is a very effective way to encourage them to practice, get feedback, revise, and so forth.

White boards can also support check ins. For example, you can write a check-in question up on the board and have students write up their responses. If you do this on the front board, you can have a whole bunch of markers (or chalk) available so that students can write, and for students who may be reluctant to write, you can just walk over to them and hand them a marker. Here are some potential prompts I would use this with (mathematical and not):

- How are you feeling today?
- What is the big idea you took away from today's lesson?
- Which problem solving strategies (or proof techniques) did you observe today?
- Is there anything you would like for me to know?

This is a quick way to get input from a lot of students that is semi-anonymous, and it gets students up and moving which is always a plus.

5.1.5.5 Virtual Structures

The default assumption of this book as that you're teaching in a face-to-face classroom, but the vast majority of what I have written can be applied to a virtual classroom (synchronous video meetings). As the COVID-19 pandemic has shown us, we can never know when we'll be in person or virtual, and hybrid modalities are also useful.

Verbal communication in a synchronous meeting is like a face-to-face whole-class discussion. The biggest advantage of virtual learning environments is that you can have parallel conversations. The chat allows for students to dialogue in parallel to the main presentation. You can even explicitly ask students to respond in the chat, which will allow you to get a lot of responses. As students respond in chat, you can then ask specific students if they would like to participate verbally. This initially provides a lower barrier of entry for students to participate (through chat), and then you can elevate the status of students by inviting them to share their competence verbally.[3] Empirically, we have found that chat can allow for many more students to participate than using verbal contributions alone.

Virtual settings provide many opportunities for anonymous participation. Tools like Google Drive, Sheets, or Jamboard can allow students to work at the same time. Padlet is another tool for collaboratively sharing work. Many of these tools allow the instructor to monitor what many groups are doing at the same time. And students can work with a higher level of anonymity, which sometimes gets new students involved in the conversation.

5.2 Inviting

In conversations about participation, we often put the onus on our students. We see it as the responsibility of the students to choose to participate (or not) and relinquish our control over who participates in our class. When students hesitate to participate in our classroom, it reflects the discipline of mathematics broadly, and the environment that we have created specifically. Fortunately, there are concrete strategies we can use to disrupt this (see Chapter 4 for ideas).

A common rationalization for students not participating is that they are introverted or shy. While this may be true for some individual students, typically instructors deploy this explanation without deeper reflection upon why certain groups of students disproportionately take up space in classroom discussions. Inequitable participation patterns are common across settings and are generally predictable – cisgender, nondisabled white men tend to talk the most in mathematics classrooms. In my research, we often observe a few very vocal students who take up an even greater proportion of participation opportunities. This is not the result of individual personality differences, but rather, racist, sexist, and ableist mathematical Discourses that privilege some students over others. When we accept these inequitable patterns unproblematically and deploy explanations such as "some students are shy," we are absolving ourselves from responsibility for addressing these inequities.

Unfortunately, some of our students have learned over many years of prior experiences that mathematics classrooms are not safe places to be involved [287, 38]. They might be judged if they give a wrong answer. A classmate might snicker at their incomplete, evolving thoughts. Students may have internalized that only certain types of people get to participate in math, and they may not belong to those categories. These are defense mechanisms that students have developed to protect themselves from toxic classroom environments they have previously been in. Many of our students carry years of educational trauma [304]. Given these prior experiences, I do not advocate forcing students to participate, and there may be a small number of students who choose never to participate. Nevertheless, there is a lot we can do to create a safer environment to invite our students into the conversation.

Assume some students may feel uncomfortable the first day that they walk into your classroom. Be mindful of prior trauma to co-construct a welcoming space. Remember that every time that you engage in negative stereotypical math classroom behaviors (e.g., calling things "trivial," lecturing at a fast pace for extended periods of time), it reinforces the status quo that students have previously experienced [287]. When we regularly call on the first student to raise their hand, we communicate to other students that their voices don't matter. This chapter is all about developing concrete strategies to promote something different.

[3]See https://dismantlingmathematics.com/2020/09/25/virtual-engagement-strategies-that-dont-require-webcams/ for other strategies you can use to support engagement in a virtual setting.

5.2.1 Asking Questions

One of your primary tools as an instructor is asking questions. Your questions serve many purposes. They communicate to your students what mathematics is about. They communicate what is valued. They communicate what types of thinking are necessary to solve a problem. They offer entry points into a problem situation that can spur deeper thinking. The questions we ask can help students get started, they can help students get unstuck, and they can push student thinking deeper. Our goal is not to do the thinking for our students, but rather, ask the right questions that facilitate the students to do the thinking for themselves.

One type of useful classroom talk that is backed up by research is called *academically productive talk* [39]. Such talk focuses on helping students recognize, build on, and respond to the thinking of the teacher and their peers. It also supports deeper student reasoning and the expression of that reasoning, which enhances learning [39, 200]. Another study of productive talk showed that student participation predicted student performance, and the teacher's support of productive talk was a significant mediator of this connection [122]. Overall, these studies provide empirical evidence for the importance of supporting our students to engage in meaningful mathematical discussions at higher levels (beyond mere recall).

Academically productive talk contrasts lower-level talk that focuses on getting to the right answer as quickly and efficiently as possible. This lower-level talk is often structured between the instructor and an individual student through an Initiate-Response-Evaluation (IRE) sequence [172]. Here's an example:

- Initiate (Teacher): What do we put here in the equation?
- Response (Student): I think that should be x^3.
- Evaluation (Teacher): Actually, it is x^4.

You've probably experienced this discussion pattern as a student, and you may use it as an instructor. In this structure, students rarely if ever respond to one another. These bite-sized conversations contain little of mathematical substance, often relegating student contributions to simple computations.

Asking more authentic questions can help us elicit deeper student thinking and promote productive talk. Good questions elicit student ideas so we can build on them and deepen student reasoning. Here are some examples of questions that you can use to elicit student ideas:

- How are you thinking about this problem?
- What are your initial ideas about this?
- What makes this problem difficult?
- Where are you getting stuck?
- Does this look like anything we've done before?
- What are you wondering about that would help you make progress?

These types of questions aim elicit student thinking without evaluation. Ultimately, we're just trying to get students to articulate their thinking out loud. This helps us get to know our students as mathematicians, and equally important, it offers them participation opportunities that go beyond providing the correct answer. This allows us to engage more of our students—especially those who have been previously disenfranchised by mathematics classrooms—and is a first step to other types of participation.

Another strategy for inviting participation is to ask students to comment on the thinking of their peers. This is easiest to implement after a short peer discussion, through a think-pair-share or turn-and-talk. These questions allow students to share ideas without taking sole ownership of them, which can lessen the fear of failure or critique. These questions are process oriented and aim to invite participation. For example:

- What was something interesting that your peer shared with you?
- Did you get any new ideas from your partner?
- Was there any place where the two of you got stuck?

- What did you and your partner wonder about?
- Can someone tell me what your group talked about?
- Did your group have a consensus, or did you have different ideas?

Once student thinking becomes public, you will need to do something with it. If you immediately respond with judgment—saying something is right or wrong—you're falling back into the IRE trap. Instead, practice active listening and show interest in student thinking. You might make eye contact, nod your head, have an open posture (not arms crossed), and acknowledge a contribution with words like (all right, yep, uh huh, okay). After the student is finished sharing, here are some ways that you could acknowledge their contribution:

- That's interesting, thanks for sharing.
- You mentioned X, and I would love if you could explain more about that.
- Could you tell me more about...?
- All right, is there anything else that you would like to add?
- I wonder how what you're saying connects to...?
- Does anyone have any questions for our speaker?
- Would anybody else like to add on?
- Can somebody else explain that idea in their own words?
- Did anyone else have a different idea?

These responses are nonjudgmental and focus on furthering the conversation. Such responses help us create a conversation rich environment.

To be clear, we don't want to leave our students with the idea that incorrect ideas are correct, and at times, we will need to directly correct our students. For example, suppose you've asked students to prove some conjecture, and a student simply provides three examples in which the conjecture was true and looks very proud of their flawed empirical argument. In many mathematics classrooms, that student would be torn apart. An instructor might say,

> No, that's not correct at all. Empirical arguments are not mathematically valid. In a college mathematics classroom, we need to have real arguments.

They might response even more curtly, "No, that argument doesn't work." These responses are factually correct and extremely effective at shutting down conversation in your classroom. That individual student will be mortified, and other students will be reticent to participate.

Let's try again. After the student provides their flawed thinking, you instead respond with "that's interesting, but how can you be sure that you've covered all of the cases?" The student responds "well, after a few cases I can be sure I see the pattern, so it follows by induction." Again, the student is sharing an incorrect idea, but because you invited further participation, you are now aware that they don't fully grasp how mathematical induction works (an idea you can return to later). Now that this student has shared two in-progress ideas, you might instead invite other students to contribute. You could ask "Does anyone else have any other ideas? I'd like to hear how everyone is thinking." It's likely that another student will respond that empirical arguments aren't mathematically valid. And if no students in your classroom arrive that conclusion, it indicates you need to help your students understand what constitutes a valid argument.

At the end of this exchange, it should still be clear that the empirical argument was flawed. However, because you first focus on eliciting ideas before moving towards a correct answer, you're not shutting students down. This allows you to learn more about what your students know and to invite further conversations. You don't need to do this every single time a student offers an incorrect idea. Sometimes you can just correct it and move on, but you can only safely do so when you've built up a classroom environment in which students are not afraid to share their wrong ideas.

Broadly speaking, how you respond to student thinking can reinforce or disrupt the idea that mathematics is about binary understandings of right or wrong. If your primary response is to say what is right, students will think that math is about getting the right answer. Instead, if you emphasize process, ways of thinking, productive missteps, and so forth, students are much more likely to see mathematics as a process and as a set of practices. This viewpoint helps students see the messiness of mathematics and supports them to engage in productive struggle and productive failure.

5.3 Deepening

Research shows that not all types of contributions are equally supportive of learning. In particular, when students explain their thinking, it helps them consolidate their ideas and leads to deeper understanding [42]. One thing to consider is that even when we start with a meaningful mathematical task, how we enact that task in discussion can make it more or less conductive to supporting learning (see Section 4.4.1). Here, I consider the different types of questions we can ask that help elicit deeper thinking (and thus enhance learning) with a given task.

In the EQUIP tool I co-created, we code questions with hierarchy of *what-how-why*, with a fourth catch-all category of *other* [230]. We use this hierarchy to help teachers develop better awareness of their question types, so that they can elicit deeper and more productive thinking from their students. Teachers typically focus on the lower-level *what* questions, and our goal is to move them towards higher-level *why* questions. The categories we chose are closely connected to research on scientific communication, which highlights how student explanations are enhanced by specific forms of teacher discourse [171] and are connected to student learning [250].

5.3.1 What Questions

Mathematics classrooms are often organized around *what* questions. A *what* question focuses on providing an answer or recalling a fact. This type of question is aligned with the idea that mathematics is about getting the right answer. Here are some examples of a *what* questions:

- "What did you get for number 4?"
- "Can someone tell me what the answer is?"
- "What is the result when you plug the numbers into the equation?
- "Do we have an example of where that's not true?"
- "Do you remember the definition of a limit?"

The more frequently that we ask *what* questions, the more we reinforce the idea that math is about getting to right answers. This de-emphasizes the learning process, which can inhibit risk taking, productive struggle, and meaningful collaboration.

Most college mathematics instructors I have observed ask primarily *what* questions (often 80% or more of their questions). Ironically, these same instructors perceive that they are asking students to explain their thinking, counter to empirical evidence. Thus, showing these data to the instructors is very eye opening. I suspect that because these instructors spend a lot of time explaining the math themselves, they believe that they are creating opportunities for their students to do the same. I work with these instructors to help them utilize other question types, with a goal of using less than 50% what questions (ideally, even fewer). Although there is no specific research for this figure, I find it to be a useful reference in my professional development work.

5.3.2 How Questions

A *how* question focuses on the process that a student follows to get to an answer. Unlike a *what* question, there usually isn't a single "correct" answer to a *how* question. Rather, *how* questions get at different types of student thinking, and therefore, benefit from the diversity viewpoints students can offer. Here are some examples:

- "How are you thinking about this problem?"
- "How did you get started?"

- "Can you tell me how you did this problem?"
- "What was your process for solving the problem?"
- "What should I do for the next step to solving the question?"
- "What did you do to get your answer?"
- "How did you get your answer?"

As the examples show, a *how* question doesn't necessarily have the word *how* in it, and it could even have the word *what* in it. The key distinction here is that these questions focus on the process of doing mathematics rather than the product.

Imagine that you have students working out a proof. After providing student work time in small groups, you may wish to have a few different student groups explain how they approached the problem. Some of them may have tried a direct proof, while others used proof by contradiction. Rather than focusing only on getting a correct proof, having students show the how allows you to compare different ways of thinking. As students integrate these different ways of thinking together, it helps them build deeper conceptual understanding. In my work with instructors, I find that *how* questions are the least frequently used type of question. By increasing our use of process-focused questions, we get to know more about how students are approaching the work, which helps us build on their thinking and provide appropriate feedback. An exciting consequence of eliciting authentic student thinking is that sometimes it will surprise us, and it can even offer new ways into thinking about problems that we had never considered before. As we help students reconcile and integrate these multiple different viewpoints, it strengthens their conceptual understanding of mathematics.

5.3.3 Why Questions

A *why* question is generally the deepest type of question we can ask, which aims to get students to provide justification or explain why something works. These questions move beyond factual recall and procedures to press students to engage in sensemaking and theory building. *Why* questions elicit student explanations in a way that deepens learning [42]. Here are some examples of *why* questions:

- "Can you explain what you figured out?"
- "Can somebody tell the class why this works?"
- "Why do you need an exponent in this equation?"
- "Who else heard what Rachel said and can explain why her strategy works?"
- "How do you know that seven is the correct answer?"
- "Why does your group think that this proof is correct?"

These questions focus on explanation and justification. In general, *why* questions are the most difficult to answer (but this is not always true). Even though many students may be able to get to a correct answer and tell you how they got there, it adds an extra level of challenge to explain why they should approach the problem that way or justify that their answer is correct. When we describe mathematical concepts to our students, we use *why*-level talk all the time. We should give them opportunities to do the same. In my observations of mathematics instructors, it is not uncommon to observe an entire class session in which not a single *why* question is asked.

Why questions are also important from a status perspective. When students correctly explain why something works, it positions them as a competent and capable mathematician. Empirically, I observe that *why* questions are disproportionately asked to students who are already privileged (because instructors assume that they will be well-positioned to answer them). In this way, even if there is a relative balance in the overall quantity of participation between student groups, if all the highest-level talk is concentrated with a particular group, that group still achieves a much higher status than other groups. For this reason, using intentional strategies to help specific students (and student groups) successfully engage with why-level reasoning in public spaces is an essential strategy to disrupting hierarchies and fixing status imbalances.

5.3.4 Other Questions

Questions that fall under the *what-how-why* continuum are your primary tools for eliciting student's mathematical ideas. Nonetheless, there are a whole variety of other types of questions that play a role in facilitating discussions. Other types of questions focus on logistics, managing participation, or getting students to ask questions. Here are some examples of *other* types of questions:

- "Do you understand?"
- "Do you agree?"
- "What questions do you have?"
- "What comments do you have for the speaker?"
- "Is there anything you want to know more about?"
- "Do you want to add anything?"
- "Can someone repeat what Brenda just said?"
- "Is it possible to do that?"

The goal of such questions is to invite and organize participation. These questions can also be used to get students to build on each other's ideas, which is an important part of the recipe for genuine exchange between students.

5.3.5 Deepening your questions

The *what-how-why* continuum is designed to help you think more intentionally about the questions that you ask your students. Using this shorthand, you can become more aware of whether you're just focusing on answers or getting to deeper thinking. It may take time to get used to asking different types of questions. One strategy that you can use is to write down questions you would like to ask in your notes or lesson plan. When you reach a certain point in the lesson, you will already know which questions you would like to ask, rather than trying to come up with them on the spot.

You can also use *what* or *how* questions as a lead into *why* questions. This general strategy is called *teacher press*. Let's suppose you were reviewing a homework set. Your opening question might be "What answers did people get for #4?" This is a *what* level question, and if you stop with asking for answers it won't be very mathematically interesting for you or your students. But this question can serve as the first of follow-ups. Suppose a student offers a particular answer. Then you could ask "How did you get to the answer?" or "How do you know if that's correct?" Now you're starting to dig deeper into process and justification. Your questioning strategies need not follow this pattern. You can also jump into the *how* and *why* directly. If you start using these types of questions regularly, your students will start to justify their answers even without your prompting, because they will begin to learn that is what is expected in your classroom.

5.4 Managing

Inviting students to participate is a first step to creating a discussion-rich environment. After we offer these invitations, we need to manage the opportunities to participate to ensure that they are equitably distributed. You can benefit from incorporating individual think time, peer discussions, and small group work before you initiate a whole-class discussion. This provides access to more students to join the conversation, and will make it easier for you to manage participation from a wide variety of students. Otherwise, it is more likely that a few students will dominate, and you'll be working very hard to get others to participate. Here are some guiding principles and strategies you can use to manage participation.

5.4.1 Be Explicit

Be explicit about how you're managing participation. Tell students that your goal is to get a variety of ideas out on the table, and that you're going to use explicit strategies to achieve that. If you want students to raise their hands, tell them why you want them to raise their hands. If you've heard too much from a small subset of students, and you want to hear from others, tell your students why it is important to have multiple voices (e.g., "there are many different ways to think about the math concepts, and when we consider how multiple different viewpoints work together, it helps us deepen our conceptual understanding.").

You can choose how much detail you'd like to share, but my personal preference is to unpack my teaching practices very explicitly. Suppose you ask a question, and no students raise their hands. That's a good opportunity to say something like:

> Hey, this is important and I'm going to give you plenty of time to think about it. I have no problem with awkward, so we can just stare at each other for a bit while you get your thoughts together.

After this statement you can wait a bit. If still nobody raises their hand, you'll need another strategy. You could say:

> Okay, it looks like you all could use some more time to think about this. Please turn to a partner and discuss for two minutes, and then I'll ask the class again.

This provides more processing time, and listening to the peer conversations may help you intentionally select students who can share their ideas (see Section 5.6).

5.4.2 Use your Body

How you position yourself in the room is a great strategy for managing participation. If you stay in one place in the front of the room the entire class session, it communicates to your students that you don't plan to interact with them. In contrast, when you move around the room, it allows you to listen to student conversations and strategically check in with certain students. Rather than facilitating all conversations from the front of the room, you can also facilitate from different places. Of course, some classrooms don't allow for much movement, and that can be a barrier to effective conversational management.

Where you place your body will draw student attention to a particular part of the classroom. Suppose that one of your small groups had an idea that you want to elevate to the class. You might begin the discussion while standing at their table. Your body placement also shows students you're listening. Perhaps you are having a whole-class discussion and one of the tables is talking loudly and disruptively about something not related. Simply moving over to their table and standing next to them is likely enough to quiet them down without needing to say anything or disrupt the larger conversation. Body placement is a subtle and effective way to help manage attention in your classroom.

5.4.3 Hand Raising

Having students raise their hands is a simple and effective way to facilitate the flow of conversation in a face-to-face classroom. I regularly observe instructors who do not use hand raising, and almost inevitably the participation patterns in their classrooms are extremely inequitable. They also tend to be dominated by one to two students who speak over all others. Simply changing their practice to effective use hand raising is a very simple change that has a dramatic impact on their participation.

A rationale that instructors give for having students shout answers out without hand raising is that it creates a casual environment in which students should feel comfortable to just say what they think. However, the reality is that most students will not feel comfortable shouting out ideas, and instead, the quickest and loudest students dominate the space. Students predictably behave according to the way that they have been socialized, and this almost certainly reinforces racial and gender stereotypes. To be clear, having students raise their hands is one of the simplest and most effective changes you can make to improve equity in your classroom, if you're not doing it already.[4]

[4]Although there may be some circumstances where you want students to participate without hand raising, it should be an exception rather than the norm.

To equitably manage participation, you will need to establish that it is your role as an instructor to call on students. In your early class sessions, almost certainly you will have students—typically students perceived as higher status—who speak up out of turn, or don't follow turn-taking norms. You will also have students who may shout out the answers at the same time as they raise their hands. These are practices you will need to disrupt immediately. If you don't establish the need to raise hands and wait before participating, your class will almost certainly be dominated by those few students. Here are some ways to help prevent a student from dominating:

- "I see that you're really interested to share, but I'd like to hear from others in the class."
- "Dan, I see that you're excited about this problem. Before you share your thinking, let's get some other ideas on the table."
- "I want to remind folks that it's important for us not to shout out the answers as soon as we have them. Everybody will process the material at different speeds, and we want to give everyone time to think. Math is more about deep thinking than speed, so I don't want to rush anybody."
- "Chris, please wait, we'll get to you in a moment. It's important to me that everyone has a chance to share in this class, so I want to give others a chance to speak."
- "We all have different ways of thinking about the math, and when we get more different ideas on the table, it helps us make connections and strengthen our own understanding. There is no one "best" way to think about any problem."

All these responses follow a similar pattern – they acknowledge the student who wants to share (not shutting them down) and emphasize the importance of waiting to hear other voices. You will probably need to use these statements a few times in the beginning of the semester until the class gets the hang of it. Don't be afraid to interrupt dominant students to create a safe classroom for everybody else. Be assertive and be kind. Once you establish the norms of turn taking in your classroom, you will have much more space to promote meaningful and equitable conversations.

5.4.4 Wait Time

A common pattern I observe is that when instructors ask for students to raise their hand, they provide little think time and instead simply call on the first student to raise their hand. In some instances, even though the instructor asked for students to raise their hands, a dominant student shouts out the answer, and the instructor does nothing to reinforce for the community agreement of raising hands. This will lead to dominant students dominating, and racial and gender inequities are exacerbated. While the instructor hasn't done anything intentionally to create inequity, they are witnessing how existing societal inequities will reproduce themselves in our classrooms, unless we actively disrupt them. And to be clear, it is the instructor's responsibility to disrupt them. Another common issue that can arise is an instructor either answers their own question or rephrases the question very quickly in a way that is confusing for students.

To broaden participation, you will need to practice your *wait time* [249]. Extensive research shows that using wait time will diversify the participants in your discussion and it also leads to more thoughtful responses from students [301]. I like to describe wait time as the "art of awkwardness." The concept of wait time is quite simple. Anytime someone in your class speaks, you wait before responding. When you ask a question to your students, you wait before calling on anyone. If a student speaks, you wait before responding to them. I suggest counting to 10 silently in your head (this will feel excruciatingly long) before calling on someone. Here's one example of how you could introduce this concept to your class:

> Look, we are working on some complicated math in this class, and it's going to take you time to think about the questions I ask. When I ask a question, I don't expect you to have an answer immediately. I want you to take some time to think deeply about the material and then share your thoughts. This might feel awkward at times. You'll see me intentionally waiting and not calling on students who might be the first to have their hand up. This is called "wait time," and research shows this is an important part of us having more meaningful and equitable conversations.

The other form of wait time (which is also important) is what happens after a student speaks. In an effort to avoid silence, it's common to respond as quickly as possible. Instead, pause and allow for processing of the student comment. Especially when students are sharing work-in-progress thinking, it might take some to make sense of their emergent ideas. This makes it easier for students to respond to one another. Waiting after students share also allows you to emphasize student comments. If a student makes an important statement, you might pause for dramatic effect. You might even ask them to say it again, because you want to make sure that students get that important point. This allows you to elevate the status of students who may be quiet or not seen as high status in the classroom, because you're giving them public space and publicly valuing their ideas.

5.4.5 Multiple Hands

The multiple hands strategy is an effective way to promote wait time. If you call on the first person to raise their hand by default, it will skew participation in your class. The multiple hands strategy makes this explicit and transparent for our students. I use the five hands strategy repeatedly with faculty in professional development, and I find it is very easy for them to implement and it has immediate benefits on diversifying student participation. To use this strategy, I tell my students that I will wait for five hands to go up before calling on someone:

> Since it's important for me to give you enough time to think about the math in this class, I'm going to wait to call on anyone until I know that many students have had enough time to think more deeply. That means I'm going to wait until I see five hands in the air before I call on anybody. After I see that many hands, I'll choose somebody to participate.

As with the other strategies, you will need to remind your students that this is a strategy you're using. In future follow ups I might say "remember, I'm going to wait for five hands before I call on anyone."

The beauty of this strategy is that after you have five hands (or however many hands you're waiting for), you can call on anybody. It might be the first person to raise their hand. It might be the fifth. Or it might be someone who didn't even raise their hand. You might say "Hey Louise, I know you didn't raise your hand, but I'm curious if you can share any thoughts on this." Louise might decline, but at least you've provided think time and made a gentle invitation. Because you're waiting for multiple hands to come up, you know that at least some portion of your class had enough time to process your question.

5.4.6 Think-Pair-Share and Turn-and-Talk

I have mentioned these strategies in passing a few times already, but they are so powerful that they deserve their own section. These strategies are slightly different but essentially serve the same function. In a think-pair-share, students have individual think time, then they talk with a partner, and then it comes back to the whole class discussion. In a turn-and-talk, the individual think time step is skipped. The length of the "think" and "pair" portions can be adjusted depending on the question you ask, but I would typically make the think time shorter (under a minute) and may have a couple of minutes for partner discussion. These discussions are intended to be short because your primary goal here is to pause a whole-class discussion and feed back into that discussion.

The benefits of these strategies have been widely studied across settings, showing improvements in student motivation, participation, and learning [85, 96, 114, 128, 197, 253, 284, 321]. The reasons these strategies work is connected to ideas we have already explored—they provide students with opportunities to share their thinking, explain ideas, receive feedback, and build identities as participants. Simultaneously, we should not take the strategies as a panacea, because as with all techniques, inequities can arise [60]. Depending on how the share portion of the think-pair-share is managed, it may lead to inequities (e.g., discussions dominated by white men) or stress (e.g., if students are cold called). Fortunately, with simple strategies we can avoid these pitfalls.

Here's an example of how to use these strategies effectively. Imagine you ask an important question that you want students to engage with. You ask the question with enthusiasm, but the response is silence (the audience is all crickets). Not a single student speaks up or raises their hand. At this point, you have a few options. You could answer your own question, but this isn't good because it discourages participation in your class. Another option would be to reframe the question. This is a good strategy sometimes, because it could be that your question was just confusing, and students didn't know what you're asking. But let's suppose you have a clear question, and students just don't have an answer or don't want to share. In that case, use the turn-and-talk strategy. [5] You say something like:

> Okay, I can see this is a deep question, and it's one that I really want to discuss. Let's do a turn-and-talk for two minutes, and then come back to the whole class discussion. So, turn to a partner for two minutes, discuss, and then I'll ask for ideas again.

This strategy does a lot. It communicates that you still want students to grapple with your initial question. It also gives students more think time. During the two minutes of partner discussion, you can walk around the room to overhear conversations and see what students are thinking. If there was something confusing about the initial question, you can usually figure out what it was. After the turn-and-talk, you can come back to the whole class discussion and get student thinking out to the original question. This will be especially effective if you use strategies like obtaining student consent to share (e.g., asking students during the pair portion if they are willing to share later, so they are not surprised or put on the spot).

5.4.7 Selecting Students

Now that we've considered a variety of strategies for managing participation, let's bring it all together to think about how you select students. Through a combination of strategies like setting explicit expectations, using your body placement, having students raise hands, using wait time, and turn-and-talk, you should be able to broaden the who volunteers to participate in your class. Ideally, this will include students from a wide variety of social marker identities, and as you use more management strategies, hopefully even students who haven't participated in the past will choose to do so. Who you choose to call on should be guided by what you know about inequities broadly, and specifically about how they have played out in your class (the best way to know this is by collecting local data, see Chapter 7). One way to support yourself to bring new students into the discussion is by including the names of students who haven't participated as much in the past into your lesson plan, so that you're reminded that they are students you hope to bring into the conversation today.

It is also possible that you may find that most students from particular minoritized identities do not wish to participate. It could be that using stronger facilitation strategies such as assigning competence (see Section 5.5) or orchestrating discussions (see Section 5.6) can remedy this. This could also be evidence that there is something with your classroom climate that needs to be shifted (see Chapter 4). It's also possible that there are some toxic elements of the classroom that need to be addressed (see Section 5.7.1).

5.5 Assigning Competence

Congratulations! If you've correctly followed the strategies for inviting, deepening, and managing participation, your classroom should be full of participation from a variety of students. At a minimum, you have provided the foundation for a classroom that is grounded in student thinking. To disrupt hierarchies, you'll want to go one step further, and think about how you draw positive attention to student ideas with intentionality. This ties into the teaching strategy called *assigning competence* (reference Section 2.3.7 on Complex Instruction). This strategy is powerful for two main reasons. First, it allows you to draw attention to authentic and meaningful mathematical contributions, which helps your students understand what matters mathematically. Second, as a part of this process, you're able to highlight that contribution as coming from a particular student, which elevates the social standing (i.e., status) of that student. As the instructor, you have more power, authority, and status than any student in the classroom. In this social space, that means that you have the capacity to give weight to some contributions over others, and in turn, you can create (and disrupt) hierarchies in the classroom.

Assigning competence has three main components: 1) identifying a student perceived as low status, 2) identifying their mathematical contribution, and 3) making that contribution public. Identifying students perceived as low status is nontrivial, but can be supported by intentional data collection (see Chapter 7).[6] Next, you will use strategies to elicit

[5] A turn-and-talk is very difficult to implement in a virtual setting, so in such a setting, you might consider using a breakout room instead.

[6] The identifying step is critical. If instead, you choose students perceived as high status who often participate, and you further elevate their contributions, you are unlikely to distrupt inequities!

participation, and then you monitor that participation until you identify a mathematical contribution from a student perceived as low status.[7] The final step is to make that contribution public, as a way to elevate the status of that student. The strategies in this section are all examples of making the contribution public.

5.5.1 Spotlight

The simplest strategy is the spotlight. The goal here is to highlight the contribution of a particular student, creating public space for it to be valued. Your goal is also to allow other students to process and learn from that contribution. You might use words like, "I want to pause for a minute while you think about that idea," or "I heard something really interesting in the partner share that I want you all to hear," or "I'd like you all to think about Alyssa's argument." In each of these examples, you use your authority as the teacher to bring attention to a student idea. To be clear, I wouldn't recommend drawing extra attention to a student's incorrect idea, because that could embarrass or demoralize them.[8]

5.5.2 Naming Ideas

This strategy can be used in conjunction with the spotlight. The goal here is to attach a student's name to a particular idea. This could be something like "Katy's conjecture," or "Kehlani's argument." This strategy is especially powerful when there's a mathematical idea, approach, or conjecture that you would like the class to focus on or return to later. I've witnessed some classrooms in which a student idea is taken up by the class for multiple days. Imagine the impact on that student's status! As with the spotlight, you will want to be sure that you use this strategy in a way that genuinely portrays a student in a positive light. Additionally, you must be aware that if you use this strategy to further bolster the status of students who already have status, you may be working against your own interests of disrupting hierarchy.

5.5.3 Teacher Press

Teacher press is the process of asking a student to explain more about their thinking. Suppose that you ask a question to the class, and a student responds with something surface level like "I got four." Such a response tells you nothing about how they got the answer, or if it makes sense. After receiving such a response, you can use follow up questions to move the student along the *what-how-why* continuum. As you spend time working with that student to get to a deeper level, you're drawing more attention to them in the social space. For a student who can portray their mathematical competence, this can be a positive light. For a student who is totally confused, they might receive this as negative attention. If you use this strategy in conjunction with what you already know about student thinking gleaned from a turn-and-talk or small group work, you can press the right students at the right time.

5.5.4 Revoice

Revoicing is the strategy of restating what a student said. Sometimes you're simply repeating the same words. "Jillian said that we need to use Sylow's theorems." Other times, you might be stating essentially the same idea, but with different words. Suppose a student Justin says something like "this won't work at a corner of the function." If the context is clear, you might "I hear Justin saying that we can't use this approach, because the corner of a function is not differentiable." After revoicing Justin's comment, you might say something like "What do you all think of that?" This is the power of revoicing. It allows you to take a student idea, make space for it publicly, and let others respond. Again, when you elevate the productive ideas of students, you can elevate their status in the classroom hierarchy.

[7] To be clear, there are some mathematical ideas that may not be valuable to highlight. It's important to wait for a moment when a student has something genuinely interesting to share about the problem at hand, so that you can share it with the class. This requires additional attention to monitor the ideas of students whose status you're hoping to elevate, so that you can use the opportunities that arise.

[8] There are exceptions to this. In a classroom environment where you have normalized failure as a productive part of learning, it is possible to have students share their incorrect ideas in a nonthreatening way, but this is a subtle act to pull off. In the case that you are using such a strategy, I would want you to be very sure that the students aren't feeling a sense of shame from what you're doing, and I would recommend using this primarily to highlight incorrect ideas from high status students who would be less likely to feel deflated from being incorrect publicly.

5.5.5 Competence

Perhaps the strongest way to draw attention to a student is to explicitly highlight competence. Highlighting student competence is a process of noticing a student's mathematical contribution and explaining why it is valuable (see Section 6.2.1 for how this idea can play out in student feedback). A key feature of focusing on competence is that it goes beyond simply noting a contribution or offering empty praise, to be highlight the specific mathematical contribution. This might sound like:

- Devon found two solutions to this problem. Having multiple solutions is a great way to deepen your understanding of the concept, so our class can all learn from this.
- Erica's hand gestures are showing us exactly how the function would behave. This gives us another useful way to look at this situation as compared to our written graphs.
- Matias found a clever way to work around the issue that was preventing us from proving this conjecture earlier. Let's listen to the idea!
- Claire's group thought of a way that we can take this problem even deeper by changing one of the assumptions. I think this is a great opportunity for our class to push our thinking.

As each of these statements makes clear, there is a specific mathematical contribution (e.g., multiple solutions, multiple representations, proving a conjecture, exploring a problem deeper) that the comment is highlighting. Such statements both draw attention to positive mathematical ideas and also elevate the students who contributed them by acknowledging the contribution with specificity.

5.6 Orchestrating

The prior sections focus primarily on facilitating discussions spontaneously. To take our discussions to the next level requires advanced planning to orchestrate discussions. There is an adage that a politician never takes an issue to vote unless they already know the outcome of that vote. By analogy, ideally, you would like to know exactly how your discussion will unfold before it begins. Drawing on the strategies in this section, you can exert a high level of control over how discussions unfold, in a way that appears very natural and organic to your class.

We will draw on a set of "five practices" developed by mathematics educators [274]. Although these practices were initially created for school mathematics, I have used them extensively in postsecondary mathematics [223]. These practices are as follows: anticipating, monitoring, selecting, sequencing, and connecting. The typical instructional sequence that leverages these practices has three parts: introduction of a task, collaborative problem solving, and whole class discussion. While these strategies tie in closely to academically productive talk and assigning competence. To make these strategies work, you observe student thinking during the collaborative problem solving, using what you learn about student thinking to organize the discussion. These strategies only work if students have time to work on a problem before the discussion. Let's walk through the steps.

5.6.1 The Five Practices

When choosing a mathematical task, you want to be able to *anticipate* how students will respond to it. This is difficult with a task you've never taught before. If you have used a task before, you can make notes from prior years about sticking points and areas of productive struggle for students. Keep these in mind before the task begins. This will help you ensure that you can highlight key mathematical ideas that come out of the problem. Or, if you are working with a high-quality math curriculum, you may already have access to samples of student thinking that can support your anticipating.

The next two strategies are used while students are working on the task. First, you will *monitor* what students are doing in small groups. Use questioning strategies to elicit student thinking. Personally, I like to carry around a clipboard so that I can make notes about what different groups are saying while they are working. Over the course of an extended problem solving session, you will quickly lose track of student ideas and which student they came from, unless you write them down. Your notes provide the foundation for planning a beautiful discussion later.[9]

As your collection of notes grows, you'll want to start *selecting* which ideas are most important to highlight during the discussion. While you're selecting ideas, you should let the students know in advance that you'll want to call on them later. This reduces stress and gives your students time to prepare. This might sound something like "Eloise, I really like the strategy that you're using here. When we come back as a whole class, would you mind sharing it with others?" This statement is doing a few things. First, you're giving Eloise an advance warning. Second, you're encouraging Eloise that her thinking is important for the class to know about. Third, you're giving Eloise a choice. If she doesn't want to share, that's okay too. If Eloise declines, you still have options. You may offer another student in that group the chance to share. Or you could ask if you could share her idea for her in the whole class discussion. Lastly, you can share the idea anonymously.

The last two strategies happen near the very end of work time. When you *sequence* the ideas, you will want to choose the order in which student ideas come out. The idea is that sequencing gives you a great deal of latitude in inviting participation opportunities that normally wouldn't exist in a non-orchestrated discussion. If the first idea that is shared out is a complete and correct closed form solution to the problem, it doesn't really allow you to get into the messiness of mathematics in the making. But if instead, you start by getting some productive attempts to the problem that may have not been fully worked out, you can give the class a whole lot more to think about. When I orchestrate discussions, my goal is to try to get at least one idea out from each group. I think carefully about what each group's work might productively offer to the conversation.

The final step happens as the ideas are coming out, which is that you help *connect* the student ideas with each other to bring together coherence in the discussion. For instance, you may have groups that used two different approaches to solve the same problem. This is a wonderful opportunity to connect different mathematical representations that describe the same situation. Someone may have used proof by contradiction where another group used induction. How do those arguments relate? Why might they both work? All those deep mathematical questions can come out in this connecting phase.

5.6.2 An Illustrative Example

Let's walk through an example of this to make it more concrete. I've published about this more extensively elsewhere [223], but here's a brief overview. The context for this example was a graduate-level course on real analysis. We were working on space-filling curves. The specific task focused on the Lebesgue curve (or Z-order curve), which can be defined as a limit of a sequence of functions over the unit square, which provides a mapping from Cantor Middle Thirds set to the unit square. To make sense of this, students had to go through some binary to decimal mappings as well (see Figure 5.1).

In this course, students were working in groups, and I let them choose the names for their groups. I had the following groups in my class (these names are relevant to my monitoring process):

- Mathletes
- Function Families
- Angles
- Cauchy's Island

To launch the task, I demonstrated how the mapping for the Lebesgue curve would work for four subdivisions of the square. I then asked students to figure out the next member of the sequence (a three-by-three division), and eventually work their way to full solution. This was my first time teaching the task, so *anticipating* was difficult. I used what I knew about the students and their backgrounds (e.g., experience with binary and decimal mappings), but I didn't really

[9] As mentioned earlier, I have also seen instructors use homework as a tool for monitoring student work to later highlight during a class discussion. This is a strategy I have never actually seen written about in the literature, but it was one that spontaneously emerged in a learning community I was running with three engineering faculty members who found it easier to email students ahead of time rather than trying to do the monitoring during class.

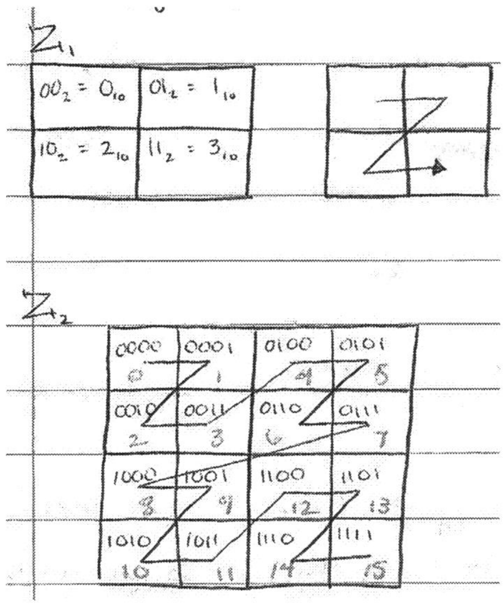

Figure 5.1. The first two iterations of the Lebesgue curve.

know exactly what to expect. Even though this is a set of five practices, I find if I haven't engaged in proper planning, I often neglect to implement the first practice very well. This especially true for courses that I only teach every few years because student thinking is just not fresh enough for me to be useful. If you plan appropriately, you could use a pre-assessment to surface what students know about the problem to help you anticipate more effectively.

During student work time I *monitored* the group work to notice the different ways that groups were approaching the binary to decimal mappings, and what patterns they were noticing. I first noticed that *Function Families* had recognized that each binary expression would increase by 1, as they traversed the curve through the squares. Thus, rather than performing the actual conversions, they just labeled the squares 0 to 15 (as shown in Figure 5.1). I noticed that other groups shortly discovered the same thing. However, the *Angles* worked out the expansions from left to right (rather than following the curve), so they were working out each conversion without noticing a larger pattern. I made a note to myself to come back to this difference in reasoning during the whole-class discussion, to highlight a general problem solving strategy.

I continued monitoring groups and noticed that the *Mathletes* were spending their time trying to draw the third-order curve without bothering to determine the binary expansions. However, without the expansions to guide them, this process was error-ridden and the group struggled to produce a correct curve. The *Function Families* recognized that the Z-pattern was recursive. Between each iteration, they could subdivide each box into fourths, and draw another Z where the previous box was. This allowed them to easily develop the third-order curve without looking at binary expansions. The final group, *Cauchy's Island* began to conceptualize the binary expansions as compass directions (see Figure 5.2. They started with the first-order curve. The first binary digit could be 0 (top-half; North) or 1 (bottom-half; South). The second binary digit indicated 0 (left-half; West) or 1 (right-half; East). Using nested subdivisions, the group generalized this strategy to the second and third-order curves.

While I was monitoring the groups and their ideas, I took notes on a clipboard. My notes looked something like the Table 5.1. To be clear, through the lesson I would fill out the Group and Strategy columns, but leave the Sequence column blank, until I decided the order of strategies that I wanted to highlight. This column was designed to support the *select* and *sequence* steps.

In this case, I decided that there were two big ideas that I wanted to come out during the whole-class debrief. First, I wanted to highlight how different groups were approaching the binary expansions (or lack-thereof, in the case of the *Mathletes*), as the opening strategy to tackling the problem. To highlight this, I wanted to first select the *Mathletes* who were skipping this, as a way to highlight how they got stuck.[10] Next, I wanted to move to the *Angles* and how they moved left to right. They were able to work out the expansions eventually, but it took a lot of time. After that, I

[10]By this point in the semester we had already set up norms around productive struggle, and this particular group had enough status that they would be okay highlighting a challenge like this.

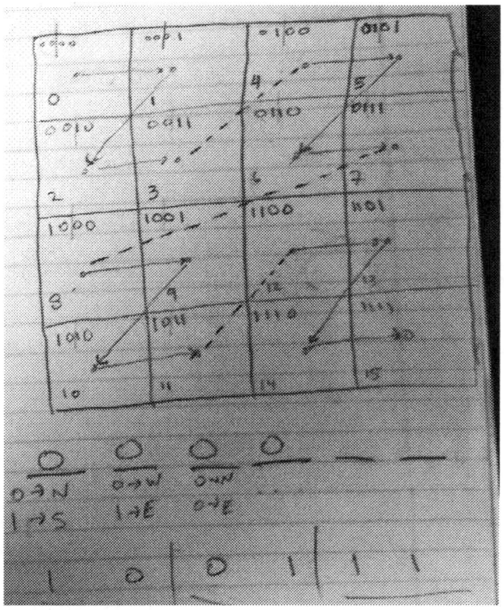

Figure 5.2. Viewing the Lebesgue curve in terms of compass directions.

wanted to contrast the *Function Families* strategy of noticing the pattern to quickly create the conversions. These were the first three steps sequenced into my debrief. By focusing on how students got started on the problem, I wanted to highlight that process matters.

Sequence	Group	Strategy
3	Function Families	Binary expansions along the curve
2	Angles	Conversions going left to right
1	Mathletes	Skipping the binary conversion
4	Function Families	Recursive Z pattern
5	Cauchy's Island	Compass directions

Table 5.1. A sketch of what my notes looked like to monitor and sequence ideas.

Next, I wanted to shift the discussion to the solutions that students were finding. Given that *Function Families* had come up with a solution, and that they would already be presenting, I sequenced their solution next. Finally, I would come to *Cauchy's Island* last, as their approach would provide another correct solution to compare and contrast. When it came to the actual discussion, I helped *connect* the ideas between the groups.

In each of the groups, I made efforts to help include students who were typically quiet in the class into the discussion. I would ask those students explicitly if they would be willing to share. When it came to sharing out the discussion, my role was to help introduce student groups in different orders. That sounded something like "I want to start with the Mathletes, because they initially got stuck on the problem, but then had some productive strategies to get unstuck." After allowing discussion and comment on that approach, I move on to the next. Eventually, I would say "Esperanza can you work through your mapping with 64 squares? Tell us more about what you found there." After more discussion, I would move to the next. "Now let's talk about the compass-directions approach from Matthew's group." After those ideas came out, I would help students connect them. "Can we see how Matthew's approach maps on to the description Esperanza gave?" As we're connecting ideas, we can also connect them to larger mathematical practices (e.g., communication) or learning goals that we have for our students.

If you think about that discussion, it is built squarely around student ideas, yet I knew almost exactly how it would go before it happened. This creates an image that all the meaningful work was governed by the students, but I had been working tirelessly behind the scenes to ensure that a very interesting and productive mathematical discussion unfolded. By intentionally choosing the students who got to share their ideas and giving them opportunities to publicly show their competence, I gave them opportunities to increase their status, which helps disrupt classroom hierarchies. To be clear, this example isn't a perfect model of this process, but it does highlight how I've used it in my own practice in a concrete way.

Notice that in this sequence I moved from less general to more general solutions throughout the sequencing. While this is often a good approach, I caution that if the same group always is the one to go last, it will still create the perception of them having the highest status. You'll want to use different techniques to mix up who gets to share and what they get to share, and make sure a wide variety of students get opportunities to share their brilliance and not just the same few.

5.6.3 Pre-Selecting Before Class

A typical implementation of the five practices relies upon work that students are doing in class. Another way to implement this approach is to pre-select to share work that they have prepared outside of class, such as on a project or homework assignment. Especially if you have had a chance to look over homework assignments before a class session, you could email a few students in advance of the class to ask them if they would be willing to share ideas from their solution so that the class can benefit from them. One advantage to this approach is that allows you to select students with great intentionality, and it requires less thinking on the fly, because you can deliberately go through student work to select specific choices, rather than trying to do in the moment of classroom engagement.

5.7 Conflict

Everything we've done so far helps create productive engagement. However, conflict is an inevitable part of social engagement. If you have students working together in your classroom, it's more a matter of when, not if, conflict will arise. Effectively dealing with conflict requires advance preparation so that you'll be ready to act appropriately in the moment. Although conflicts can arise for any reason, from a perspective of disrupting hierarchies, we need to pay special attention to negative interactions that are grounded in students' identities. Specifically, students who are minoritized due to their race, gender, disability status, sexuality, and so forth, face unique forms of threats that non-minoritized students do not. In order to address these issues, instructors first need to be aware of them. Building relationships with students is a great first step. Also, having ways to check in with students both informally (i.e., before and after class) and formally (e.g., through feedback surveys) can help diagnose issues.

5.7.1 Toxic Classroom Environments

A large body of research documents how racism [168], cisheteropatriarchy [59], and ableism [195] play out at the classroom level, resulting in identity-based threats and attacks to minoritized students. By default, a mathematics classroom can be seen as an unsafe space for minoritized learners. This places a multitude of burdens on minoritized students, both due to the weight of anticipation of negative events, and then the actual processing of events as they do happen. For this reason, it is especially important for us to build trust and community in our classroom. Equally important, we need to be prepared to disrupt these identity-based attacks as they unfold.

5.7.2 Responding to Problematic Interactions

What should you do when a negative identity-based attack comes up in your classroom? There is no "right" answer on how to respond. But perhaps there is a wrong answer. If problematic interactions arise in your classroom and you do nothing to address them, then implicitly, you are sanctioning what just happened. As a general guideline, it's important to address an interaction in the same venue that it took place. If a student made a racist comment in a whole class discussion "you got it right because you're Asian!" you should address the remark in that whole class discussion. If a student makes a sexist remark in a small group interaction ("women can't do math"), then it should be addressed within that small group. And if a student shares something ableist that happened with you after class ("someone called me retarded because of my suggestion"), that could be addressed through a one-on-one check-in with the student who made the comment. When students mistreat each other, it impacts both the target and all other students who were present, because it signals to them that this is a toxic environment. As the instructor in a classroom, it's your responsibility to interrupt these types of behavior.

Why should you handle instances publicly? Because if you don't, it will send the message to your students that you tolerate intolerance. Mathematics classrooms are cultural spaces in which student identities matter. If there are identity threats in your classroom, you need to handle them as a serious manner. In instances where the problematic comment is blatant, you can simply refer to your community agreements "This is an environment for all students to learn, and racist speech is not acceptable here." Or "I heard someone say that Niral should solve the problem because he is Asian. While you may have meant that as a joke, you have to understand that such comments can be harmful and reinforce stereotypes."

5.7.2.1 Microintervention Strategies

Anytime an identity-based threat arises, you must make a judgment call on how you want to respond to it. One resource you can draw upon is the set of microintervention strategies developed by Sue and colleagues [288]. The strategy you implement depends on your strategic goal. Are you trying to make the invisible visible? Or disarm the microaggression? Do you aim to educate the offender? Or do you need external intervention? The approach you take depends on the incident.

Suppose somebody frequently talks to their peers by only saying "you guys," or when they get an answer that they don't expect to a problem they say, "that is totally crazy!" These are forms of sexist and ableist language that have been so normalized you may not even recognize them as problematic. In instances such as these, an intervention to make these visible, saying "what do *you all* think? (emphasis on you all) or "I wouldn't call that crazy, but it is interesting." Such a response is subtle, it doesn't draw excessive attention to the original comment, but it also communicates that you did hear the comment and it's not the type of language you would use in this classroom. In these cases, the speaker may be well-intentioned and not even realize the implications of their statement. This is exactly the reason for pointing something like this out, so folks can work on finding better words to use in these situations. This can also serve as the entry point to a larger conversation, connecting back to community agreements, or even systems of oppression in mathematics.

Other situations will warrant a stronger response. Let's return to the earlier example where Raul speaks about women belonging in the kitchen (see Section 3.3.3.1). Such a statement goes well beyond the casual use of sexist or ableist language, and as such, warrants a stronger response. Here your goal would be to disrupt the behavior. Possible responses could include "that's sexist," "it's not okay to say that" or invoke the community standards, "this classroom is a place of respect for all students here, and sexism is not tolerated here." These responses put a stake in the ground that such behavior will not be tolerated. Most likely, the offender will become defensive "it was just a joke, not a big deal," or try to justify their behavior "oh, but Carla said…" In either case, you should be prepared to calmly repeat your stance that jokes of this nature are not funny, or it doesn't matter what Carla said, sexism is not welcome in your classroom.[11]

In addition to making a problematic comment visible, you may wish to spend time educating the offender about their statement. This is something that could happen in the moment, or you might pull the person aside when you have a moment. Typically, I find it is a useful strategy to first ask that person why they made that comment, or what they meant by it. Taking an inquiry mindset opens a conversation to uncover some of their hidden assumptions, which you can problematize. In other cases, you might need to be explicit about the ideas that they are invoking. You may choose to engage in this inquiry process one-on-one with a student, as not to pull other students into an extended conversation. Your questions might sound like "When you act surprised that Jamal was in our classroom, you were invoking problematic racial stereotypes about Black students." Or, "When you say Asians are good at math, what about people who aren't Asians? And how would that make Asian students who are struggling feel?" Or "Do you really mean all Asians? Isn't it problematic to put so many diverse peoples in a single box?" Or, "When you make homophobic comments, you're creating a hostile environment for queer students."

Finally, some situations may be well beyond what you can handle in the moment, and you'll need to bring in external help. Many years ago, I was teaching a mathematics course at a community college, and two students started shouting expletives at each other during small group time. It looked as though a fight was about to break out. In that moment, my overriding goal was to ensure that my classroom remained a safe space, and to keep things focused for the rest

[11] The MAA offered a webinar series on being an active bystander that provides some helpful strategies, http://info.maa.org/pages/1780913/23513.

of my students. My response was immediate, to take both of those students out of the classroom and separate them. I told them that they were done for the day, and that we would talk about this later. I waited for both students to leave (going separate ways), and then I got back to my class to finish the session. I later learned that these two students had an ongoing feud of sorts, and it just happened to hit a climax while they were in my classroom. An instance like this is one that warrants seeking external support from your campus administration to handle these issues and ensure that everyone stays safe.

5.7.2.2 RAVEN

Luke Wood and Frank Harris III have developed the RAVEN approach to disrupting microaggressions [106]. RAVEN is a learning-oriented tool for responding to microaggressions, which makes it well-suited for use in a college classroom. The RAVEN approach proceeds through five steps:

1. *Redirect.* Interrupt the interaction so that no further harm can be caused. In a public space this could sound like "let's pause for a minute so we can talk about the language we're using here."
2. *Ask* probing questions. Asking questions is a way to keep an inquiry-framing. You might say "I heard you say Asians are good at math. What do you mean by that?" Or, "I heard you say that Latinos are not hard working. I want to be sure that I understood you correctly. Can you clarify?"
3. *Values* clarification. This step puts the microaggression in conversation with the community agreements established by your class. This might sound like "we came up with a set of community agreements that include mutual respect. This doesn't sound respectful to me." You might also say "it's my job to ensure that this is a safe learning environment for all students, and right now you're making this an unsafe environment."
4. *Emphasize* your own thoughts and feelings. This is a place for "I" statements. Even if the microaggression wasn't directed at you, you should communicate how it made you feel. "I was disappointed to hear you don't think that everyone can do well in a math class. I think that other students would be hurt by your statement."
5. *Next* steps. To close this cycle, the final step involves a suggestion for what the aggressor could do differently in the future. This could include an apology to the students who were harmed and a commitment to not do it again. It could also involve a promise of self-education so that the person is better informed about their actions in the future.

The RAVEN can be a useful tool because it is a scripted technique that you can practice. As you work with a tool like this, you know that you have a concrete strategy in your back pocket that you are ready to deploy as needed. In addition, it aims to understand, educate, and diffuse the conflict situation, rather than ignoring or escalating it. No technique is perfect for all situations, but this strategy is a useful one to have in your toolbox to deploy as needed.

5.8 Reflection

Now that you have reached the end of this chapter, here are some questions and activities to support your further reflection. These questions will be most effective if you consider them in collaboration with others, who will have their own and different responses.

1. Think back to the guiding question of the chapter: What is one practice that you would like to commit to trying the next time you are in the classroom teaching students?
2. Thinking back to your experiences as a student, try to think of a really productive interaction, and a really unproductive one in the classroom. What were the qualities of those interactions?
3. What was your worst experience of group work? What made it bad, and how could you have made it better?
4. What was your best experience of group work? Why did it work well, and how can you recreate that experience for your students?
5. What was a moment where you felt very proud of yourself in math class? What do you remember about that situation?

6. What types of questions do you typically ask your students? How do you plan the questions you're going to ask before a lesson?
7. What was the last time that you really listened to someone for understanding (in your personal life)? What were the qualities of that interaction? How did you show active listening, and how were you responsive to the other person?
8. Discuss with colleagues some examples of when microaggressions have come up in the workplace, and practice responding to those scenarios using a technique like the RAVEN.
9. Think about the last time you were in a difficult social situation (be sure to take care of yourself and beware of triggers). This may have happened to you, or you may have been a witness to this. Think about this interaction using the Social Change Ecosystem Map (Click for Link). Which role did you play in that situation? Would you change anything? Aspirationally, which roles would you like to play if something like this would arise in your classroom? How can you get to that space where you can enact those roles?
10. Looking at the references of this chapter, what is one reading that is new to you that you would like to read in its entirety?

6

Assessment

Assessment is one of the primary mechanisms through which we communicate "what counts" in a mathematics classroom. Even if an instructor creates opportunities for authentic mathematical engagement, if assessment is narrowly defined as high-pressure computational timed exams, students will receive the message that mathematics is about executing procedures as quickly and accurately as possible. It follows that the quality of our assessments is closely related to the quality of tasks we choose. To use assessment equitably, we want to choose meaningful problems and ensure that linguistic, cultural, and other elements of the tasks are accessible to all students (see Section 4.4.1 for information on how to choose good tasks to use for assessment). As we provide a wider variety of opportunities for students to express their mathematical understanding, it expands the very notion of what it means to understand mathematics.

In the learning cycle of instruction-practice-feedback, assessment allows students to receive feedback on their learning. Often, assessment is taken as a synonym for grading. However, grading and feedback are distinct, and in this chapter I argue these two processes should be disentangled as much as possible. Assessment allows us to surface student thinking and to generate feedback for students. It helps us learn more about our students, and it helps students learn more about themselves. Feedback is often seen as something that happens after students engage in a learning process, which is too late to impact learning. Instead, effective assessments should incorporate feedback throughout the process. When feedback is incorporated throughout a learning cycle, it becomes a tool to promote self-reflection and deeper learning [25].

In contrast, grading is primarily a mechanism to create and sustain hierarchies.[1] In K-12 schools, grades sort students into different mathematical tracks, constraining the learning opportunities available to some students. In higher education, admissions exams such as the ACT/SAT and the GRE determine who can pursue further education and where they can pursue it. Mathematics placement exams either place students "on track" to complete their degrees on time or can put students into a remedial track. In this way, mathematics becomes one of the greatest barriers for some students who have STEM career aspirations. Because mathematics is seen as a signifier of intelligence, mathematics plays a unique role in the testing regimes of the US that sort students and constrain future opportunities. Although grades are viewed as a signifier of merit, in fact, they mired by complex intersections of race, social class, disability, language, and other identities.

Before you read this chapter, I'd like you to take a minute to write down some of your assessment practices. What types of assignments do you offer, and why did you choose those? How do you know if your students actually understand the material? What are common assumptions that we as mathematicians make about assessments, and could some of those be false? Please take out a piece of paper (or use an electronic device) and take five minutes to write these things down. This extra processing time will make it easier for you to see what is new to you. In addition, this creates an opportunity for you to write down *new ideas* in this chapter as you encounter them. As you read through this chapter, I want you to reflect on the following question: *Based on the evidence in this chapter about what types of feedback support learning, are you willing to change your course policies and grading policies?*

[1] Consider, for example, the ways that typical testing regimes and the system of accommodations exclude disabled students [192].

6.1 Assessment for Learning

Assessment is the primary mechanism through which feedback can be generated for students. Effective assessments help surface what students do and do not understand, how they are thinking, and where they are getting stuck. This provides information that students can use to improve their learning, and it provides information to instructors to adjust their instruction [65]. Assessment allows us to use student thinking as the foundation of creating a productive learning environment. This type of assessment is typically called *assessment for learning*, or *formative assessment.* (I address summative assessment in Section 6.4).

Broadly conceived, formative assessment practices have a largely positive impact on learning [19, 145]. For example, a recent meta-analysis found an small to medium effect size for formative assessment practices ($d = 0.34$, pooled from 126 effect sizes) [150].[2] The effect of student-initated formative assessment was even larger ($d = 0.61$).[3] This highlights the value of peer- and self-assessment, the focus of Section 6.3.

If instruction is conceived of as a process of helping a learner get from where they are to where they want to go [20], then effective instruction has to be responsive to where students are in their current thinking. Every student is different, and every group of students is different. Instructors need effective ways of understanding what students know to adapt instruction accordingly. Teaching without knowledge of student thinking simply isn't good teaching. Students also need knowledge of their current understanding, so they can adapt how they approach the learning process.

Implicitly, all instructors have their own goals for student learning. Increasingly, institutions require instructors to explicitly state learning goals on a syllabus.[4] This can be a useful practice, if in fact our instruction aims to help students reach those goals, and we choose activities aligned to those goals. In this case, the goals can help provide the "where" that the student is trying to go to, if this is communicated with clarity, the student agrees with those goals, and assessments authentically reflect those goals (rather than undermining them, for example, by focusing only on procedural computations). However, far too often, learning goals are simply put on a syllabus as lip service to an institutional policy and they play little role in the teaching and learning process.

By implementing formative assessment strategies, we can help students get abundant quality feedback, about what they are learning, so that they can continue to revise and improve towards those goals. Some feedback may come from us, and other feedback can be generated by peers or by individual students themselves. Here are some simple formative assessment strategies that you can utilize.

6.1.1 Questioning Strategies

Almost all of the sample questions provided in Chapter 5 can be used for formative assessment purposes. When we ask questions with attention to student thinking, every single classroom discussion that we have can provide a window into where our students are. Rather than conceptualizing questions as a mechanism to get to the right answer, think about yourself as a detective who is using questions to get to the bottom of student understanding. This means that you need to ask questions to lots of students (not just a few), to know what your class knows overall. Asking appropriate questions also helps students identify their own understandings and areas for growth.

6.1.2 Rough Drafts

Authentic mathematics is messy, full of false starts and missteps. To really understand what your students are thinking and how they are grappling with mathematics, you will want to surface their in-progress ideas. As mentioned in Section 5.1.5.4, using whiteboards is one productive way to get students to express their ideas publicly before they are fully formed. You can also have students bring rough drafts of their work on homework problems to class before turning in their homework (see Section 6.3.2), and structure classroom conversations around partially correct work.

The use of exemplars allows students to see samples of what a solution to a problem could look like. Exemplars are a powerful tool to support self-assessment, because as students begin to understand what variations in quality look

[2]There is some challenge in determining an actual effect size from formative assessments, given variations in definition and study design. For example, one attempted meta-analysis of 42 effect sizes of formative assessment studies in K-12 settings (19 were in math) found a much smaller weighted mean effect size of $d = 0.20$ [133]. However, the methods of this study were widely criticized [28, 170].

[3]Another, more targeted meta-analysis of peer-assessment found a smaller but notable effect ($g = 0.291$, pooled from 134 effect sizes) [145].

[4]For example, this has now become a requirement at SDSU, where I teach.

like, it helps them develop a sense of high-quality [252]. One strategy is to gather anonymous samples of work from your class or from prior semesters that help students unpack the thinking that goes on in rough drafts. Depending on your instructional materials, you may even have access to sample solutions that students can grapple with. Research shows that students benefit most from the exposure to exemplars when they have first produced their own draft [299].

Students should receive feedback on their in-progress ideas. Feedback on a finished product provides no opportunity for revision, so there is little to learn from the feedback. You can structure opportunities for revision through peer-feedback, assignment revisions, test corrections, etc. These strategies help you build a culture of revision in your classroom [125]. As your students get comfortable working through incomplete ideas, you can promote productive struggle and mitigate pressure for perfectionism. If students are afraid of judgment, they won't share their ideas, and you will know little about what they are thinking.

6.1.3 Opening Problems

I like to structure my mathematics content courses by beginning the day with an opening problem. This is a short problem for students to work on for a few minutes as they come into class. The opening problem might cover an idea from a prior lesson, or it could be the entry point to the day's lesson. The opening problem serves several purposes. First, it gives students a logical transition into "math mode," which could be especially important for your freshmen students who might have five classes in a day that all focus on different disciplines. Second, it gives you an easy way to see what your students know before you dig into the heart of your lesson. This is a form of diagnostic assessment that can inform your instruction later in that day. This provides you with much needed information to see if you need to adjust your course of action, or if you can move along with your lesson as planned. Opening problems are an effective way to address the *anticipating* stage of orchestrating a discussion (see Section 5.6).Third, opening problems create another opportunity for students to engage and get feedback, which helps their learning. This helps build a stronger learning community and interdependence between students.

6.1.4 Exit Tickets

An exit ticket complements the opening problem. It's a short task that students complete at the end of your class session just before they leave. The exit ticket can serve a lot of purposes to help you see how your lesson was received. You might ask a few conceptual questions or have students complete a short problem related to the lesson to see if they made sense of it. You can also use exit tickets to support reflective writing. Here are some questions I have used in the past to structure this reflection:

1. What was a key takeaway from today?
2. What questions do you have?
3. Is there anything else I should know?

Questions like this can provide useful insights and feedback about your lesson.

6.1.5 Other Strategies

Many of the strategies described earlier in the book can also be used very effectively for formative assessment purposes. These include think-pair-share, gallery walks, paper toss, student presentations, everyone shares strategies, and so forth (see Section 5.1.5). Essentially, anything that you can do to surface rich student thinking during your class session becomes a formative assessment when you use that information to revise your instruction appropriately. This information can also be taken up by students to gauge their own understanding and adjust their learning processes. Similarly, anything that you gather from a formal course assessment can be used formatively. In this way, formative assessment is just as much about your disposition towards using student thinking as it is an actual set of practices.

6.2 Feedback

Feedback is a key part of the learning cycle. Not surprisingly there are numerous books [29] and review articles [270, 318] written on this very topic. I take an expansive and agentive view of feedback:

Feedback is a process in which learners make sense of information about their performance and use it to enhance the quality of their work or learning strategies [110, p. 1402].

This definition centers learners as active agents in the feedback process. It is also open enough to acknowledge that the tasks we choose and the learning environment we build can also generate feedback for students throughout the learning process.

Broadly speaking, feedback enhances learning and is an important way to support our students. For example, two recent meta-analyses of feedback show medium to large effect sizes for improving learning outcomes ($d = 0.41$, pooled from 607 effect sizes) [134], and ($d = 0.55$, pooled from 994 effect sizes) [318]. Given these notable effect sizes, one would expect that more feedback is always better. However, these meta-analyses found that anywhere from 20%–30% of the effect sizes that they aggregated were actually negative (meaning that feedback *reduced* learning in those studies). This may seem perplexing, but it highlights the contextual nature of feedback. When we think of effective feedback, we must ask: effective for whom, for what purpose, and under what circumstances?

Consider the context of feedback. Any feedback we give to our students is mediated by our relationship with our students. If our students see us as cold, judgmental, and critical, that will influence the way they interpret our feedback. In contrast, if they see us as caring and supportive, our feedback will land differently. Similarly, our feedback may land differently for students depending on their identities and their prior experiences in mathematics [55, 325]. The work of Henderson and colleagues [110, p. 1406] provides one set of guidelines that can help us create an environment that uses feedback effectively:

- Capacity for feedback
    1. Learners and educators understand and value feedback
    2. Learners are active in the feedback process
    3. Educators seek and use evidence to plan and judge effectiveness of their practices
    4. Learners and educators have access to appropriate space and technology
- Designs for feedback
    1. Information provided is usable and learners know how to use it
    2. It is tailored to meet the different needs of learners
    3. A variety of sources and modes are used as appropriate
    4. Learning outcomes of multiple tasks are aligned
- Culture for feedback
    1. It is a valued and visible enterprise at all levels
    2. There are processes in place to ensure consistency and quality
    3. Leaders and educators ensure continuity of vision and commitment
    4. Educators have flexibility to deploy resources to best effect

As this list makes clear, there are a wide variety of decisions to make. Thus, your course policies will have a big impact on how feedback is taken up in your course. Suppose you want to give your students an exam to measure their understanding of abstract algebra. In the first case, you give a high-stakes timed exam. You collect student exams, assign grades, and provide written comments. In this situation, the students have already taken the exam and received a grade, so they are likely to ignore your comments, and the feedback has little value. In contrast, if you give an exam, and provide the very same feedback without a grade, and allow students to revise, they will take up the feedback in an entirely different way. You could then later assign a grade to the revised exam. This allows you to turn the examination process into a learning opportunity for students. This is just one example, but ideally, students should have regular opportunities to receive feedback and revise their performance.

To help you make sense of the complexities of giving useful feedback to your student, I have created a *Feedback Matrix* for you (see Figure 6.1). The feedback matrix contrasts four categories of feedback you can offer students, based on the information density (high or low) and the direction (positive or negative) of the feedback. For example the *Growth* category is high in information density and negative in direction, which means that you are providing constructive ways for students to improve their work. In contrast, the *Competence* category provides concrete information about what students are doing well. Understanding both what is going well and where to improve are important parts of effective feedback. In contrast, *Praise* and *Grades* both contain little information, and these types of feedback can actually inhibit learning. I now provide more information about each type of feedback and its impact.

Information Density

Direction		**High**	**Low**
	Positive	Competence	Praise High Grades*
	Negative	Growth	Low Grades*

Figure 6.1. The Feedback Matrix. Competence and Growth enhance learning, while Praise and Grades inhibit learning.

6.2.1 Competence

Competence feedback describes a situation in which we provide students with detailed information about what they are doing and why it is valuable.[5] Competence feedback relates to the idea of assigning competence from Complex Instruction (see Section 2.3.7). In that situation, the teacher identifies a student perceived as low status, identifies a contribution, and makes the value of that contribution public. The types of statements that a teacher uses to draw attention to the contribution are the same types of statements we can use to provide competence feedback. Depending on whether the feedback is public, it may play a different role in how it positions a student in the classroom. Here are some few examples of competence feedback:

- I noticed that you used the theorems we just discussed in class. This seems like a productive strategy.
- I appreciate the clarity of your error calculation. This would make it easy for anyone else to follow your steps.
- Your choice to use the contrapositive argument here was a unique choice, and it seems to have worked very effectively!
- Your inclusion of your diagrams alongside the formal proof made it much easier to follow your reasoning.
- You have a clever approach to this proof. I'd love to see you work it out to completion.

Each of these examples has a similar format. The feedback identifies a specific strength of the work, and explains exactly why it is a strength. By providing this type of feedback, it both lets the student know that they are doing something well, and it lets them know exactly what their mathematical contribution was. This does important work in supporting the confidence of students, positioning them as mathematicians, and guiding them for things they can continue to do in future work.

6.2.2 Praise

In contrast to competence feedback, a very common strategy is to give students praise (often called empty praise). Recall that meta-analyses of feedback find that about 20%-30% of feedback is actually detrimental to learning [134, 318]. The most common type of feedback that inhibits learning is praise. Often, instructors will praise students either

[5]The framing of competence feedback was first shared with me by Charles Wilkes II, and the examples in this section are based on his work with Deborah Ball in elementary mathematics classrooms. I have adapted the ideas here to a postsecondary context.

for doing well on a problem, or as a way to encourage them when they are getting stuck on something difficult. The problem with praise, is that rather than focusing the learner on the task and how to do better, it focuses the learner on themselves, and looking good in front of their peers or their instructor [107]. This phenomenon was first observed and described in terms of a student's goal orientation. A student focused on *learning* will seek out challenges to support better understanding of the material, but a student focused on *performance* is more concerned with external perceptions [77]. When we praise our students, we shift their focus from learning to performance, which consequently makes them learn less. Here are common examples of praise that instructors might give students, but are actually harmful:

- Awesome!
- Well done!
- Way to go!
- You got it!
- Exactly!
- You're a math genius!
- You're so smart at math!

Personally, I find that it is very ingrained in our culture and in my own behaviors to act enthusiastically and provide praise to students. In fact, I have to consciously stop myself from doing so, because I know that research is very clear that this practice is detrimental. If you're like me, one way curb this issue is to ensure that anytime you offer praise, it is given in conjunction with competence feedback [315]. This might sound like *Wow! That was a very productive use of the theorems we were using during the last class period. I love that you thought to go back in your notes to use those. Wonderful!* This statement includes some praise, but the praise is not empty because the overall focus is still on the student's competence. To be clear, the praise isn't adding any learning benefit in this situation, and whether the praise reduces any of the benefit of the competence feedback is still in need of more empirical study.

Research on goal orientation and praise has become popularized through the concept of mindset [76]. Mindset refers to whether one believes that intelligence is innate or malleable. When students believe that people are innately good at math or not, and that can't be changed (i.e., a fixed mindset), it inhibits their performance. In contrast, a *growth mindset* refers to a set of beliefs around the malleability of intelligence and improving learning through effort. Overall, a growth mindset is more associated with learning goals. In contrast, a focus on performance (e.g., through praise) can shift students towards a fixed mindset. Especially if students continue to see the same students praised, it reifies the idea that some people are just smart at math while others are not. Fostering a growth mindset requires a focus on effort, and might sound like:

- I noticed that you didn't give up on this problem, even when you got stuck.
- I can see that you're working hard at this.
- I heard you say you can't do this problem. While you may not be able to do it yet, keep working and I am sure you'll get there.
- Can you tell me what you're grappling with here? I can see you're working hard, and I know you're going to learn a lot.
- Let's keep practicing these types of problems until you feel totally confident with them. I know you'll get there.

These are just a few examples and are statements you would most likely make during small group work time. Overall, the commonality is that you are framing the learning process in terms of hard work, effort, and persistence. These statements emphasize the malleability of intelligence. This contrasts statements like "this problem is trivial" or "you are a genius," which highlight a static view of intelligence.

Research shows that simply changing mindset beliefs (without teaching math content any differently) can help students learn math better. This was demonstrated in an experimental study of seventh grade students ($N = 95$); in the experimental condition students participated in eight intervention sessions of 25 minutes to promote a growth mindset, while the comparison group learned study skills [21]. The authors found that one semester later, the average GPA of students in the experimental section was 0.3 points higher. Overall, mindset studies have been replicated across a wide variety of contexts, and the results generally hold up, but with heterogeneity depending on the context [324]. And to be clear, mindset can be taken up in problematic ways when societal inequities are attributed to student beliefs and lack of hard work. But if we embed the concept of mindset within a broader understanding of systems of oppression, it can also support learning.

6.2.3 Growth

Growth feedback is a type of feedback that provides specific information about areas to improve on a task. Not surprisingly, meta-analyses of feedback show that when learners are provided with information that helps them learn how to do a task better, their performance is improved [134, 318]. In contrast to praise, effective feedback focuses the learner on the task [107], rather than external perceptions. In accordance with the questioning strategies from Section 5.3, useful feedback could tell students about the *what* of problem solving (getting the right answer), in addition to the *how* and *why*. Feedback that focuses on *how* and *why* can be especially productive, because it attends to lifelong learning skills across contexts.

Growth feedback is critical, so it may not always be taken up in a positive way. Especially when we consider differences between our identities and those of our students, legacies of math trauma, and the pressure to perform well, growth feedback may be interpreted as simply critical and demoralizing, rather than supportive. To address this challenge, we can learn from work on *wise feedback* [55, 325]. Wise feedback is a specific form of critical feedback that has three features:

1. Specific critical feedback.
2. Explicit invocation of high standards.
3. Assurance that that the student is capable of meeting the standards.

Wise feedback was developed as a remedy to the mentor's dilemma [55], when a mentor (especially from a majority group) is providing critical feedback to a minoritized student who is negatively stereotyped. Faced by this dilemma, individuals may simply withhold critical feedback because they are worried about invoking negative stereotypes or being seen as not supportive towards their mentees. Yet, this robs the very same students of useful feedback that could support their growth. The wise feedback is a specific intervention that helps moderate the negative effects of the mentor's dilemma. Here are some samples of what the wise intervention might sound like:

- I'm giving you this feedback because I have very high standards, and I know that you can meet them.
- These problems are harder than ones we've worked on in the past. I know these are a challenge, and you're not quite there yet. But I believe you will get there, and here's some feedback to help you.
- Creating a flawless mathematical argument isn't easy, but it's a good way to push yourself. You have some very productive ideas with your initial draft, and if you take my feedback into account, I'm sure you can do even better.
- I know you've worked hard on this, so I want to give you honest feedback. Your work is good enough to pass the course, but I believe that you can reach a higher standard that we as mathematicians hold ourselves to in peer review.

In addition to the framing offered by the wise intervention, we still want to provide specific feedback about the areas for improvement. This would sound very similar to competence feedback, except it is focused on shortcomings rather than strengths:

- You choose a contrapositive argument and that didn't quite work out. Could you take another approach?
- There are some gaps in your argument, specifically in lines 3 and 5. Can you fill out the details so that your argument follows?
- I can't really follow your reasoning here. Perhaps a diagram would help make it clearer?

- I wonder if you could use the theorems we discussed in class. Starting from scratch seems like it's making this much harder for you than it needs to be.

To be clear, writing all of these comments out, including the wise interventions is a lot. To give useful feedback and to keep it manageable, we have to consider when and how to give feedback. Additionally, we can leverage peer- and self-assessment as other ways to generate feedback (see Section 6.3).

6.2.4 Grades

The final form of feedback in the matrix is grades. Depending on whether grades are or high or low, they may have a positive or negative direction. In either case, grades are still detrimental to the learning process. Like with praise, grades focus a learner on performance, rather than learning, and as a result, reduce intrinsic motivation [33]. Even if grades are high, because they focus the learner on performance, they can inhibit future learning. Because grades are often given in high-stakes situations and they determine future career aspects of our students, they have a very powerful impact on shaping behaviors, even moreso than other types of feedback. In fact, research shows that while written comments can enhance learning, when they are paired with grades, the grades actually undermine and eliminate the learning benefit of the written comments [135].[6] Because grades eliminate the learning benefits of feedback, it is important to decouple grades and feedback in the learning process (see Section 6.5 for alternatives to traditional grading).

6.3 Peer- and Self-Assessment

One way to activate students in the assessment process is through peer- and self-assessment. This helps foster independence and interdependence between students. Additionally, it provides an abundant source of feedback that exceeds what you can provide alone as an instructor. Even though student peers may not share your same level of mathematical expertise, they can offer useful perspectives to their classmates, and overall, peer feedback will enhance student performance [145, 219].

Broadly, how we organize assessment activities in our classroom communicates what we value. When the instructor is the only source of feedback and assessment, the instructor is positioned as the sole mathematical authority. In contrast, through activities such as peer-assessment, we can build positive interdependence between students and activate them as resources for each other's learning. Not surprisingly, this improves learning. Moreover, this type of community-based learning environment helps disrupt the idea that mathematics is individual and competitive.

6.3.1 The Assessment Cycle

The assessment cycle is one framework that I have developed to help describe the myriad learning opportunities embedded in peer assessment [220]. Students can learn through a variety of activities, such as: analyzing peer work, providing feedback to peers, receiving feedback, conferencing with peers about their work, and revising their own work. Each step in this cycle affords different types of learning opportunities. For example, analyzing peer work and providing peer feedback helps students to develop a concept of high quality work and to create a critical lens that they can later apply to improve the quality of their own work. Similarly, revision allows students to take input from others to strengthen and improve their own work. All these steps help students build their feedback literacies, in a variety of different ways.

More broadly, peer assessment helps create positive interdependence between students in your classroom and helps them develop skills to monitor their own learning. These lifelong learning skills will benefit your students greatly as they move on to future courses and eventually into industry. The assessment cycle is designed to leverage both peer-assessment for community learning, and to develop assessment literacies through peer-assessment that later support self-assessment.

[6] In more detail, a recent meta-analysis comparing grades and written feedback found that students received written feedback without a grade, they significantly outperformed students who received a grade with a large effect size ($d = 1.14$, pooled from 71 effect sizes) [135]. Notably, when students received grades and comments together, their performance was not significantly different from students who received grades alone (i.e., the presence of the grades eliminated the benefits of the written comments). Similarly, students who received comments alone had higher intrinsic motivation than students who received grades and comments or grades alone ($d = 1.60$).

One barrier to self-assessment is a well-studied phenomenon called the Dunning-Kruger effect [74]. This effect describes a double-bind, in which people who are struggling in a content area are unable to recognize their own struggles. As a result, people who are struggling with mathematics concepts may not realize that they are not understanding them and fail to take appropriate action to improve their understanding. In contrast, students who have a deeper grasp of the material are better able to see the limitations of their own understanding. In this way, the skills of performance and self-assessment are at least partially related. Similarly, I would argue that one of the most important things mathematics students learn through their graduate education is how to learn math, in part because they have developed enough sophistication to better recognize what they don't understand and how to remedy that.

If we want to teach students how to improve their self-assessments, it helps to unpack the components of self-assessment. Self-assessment is a complex and iterative process that includes several interrelated steps: understanding learning criteria, gathering feedback, and reflecting on one's skills relative to the feedback [323]. Thus, to learn to effectively self-assess, students need to develop proficiency with each of these component skills. For instance, even if a student is adept at gathering feedback, they may lack the judgment skills to distinguish which types of feedback are useful and which are irrelevant. These feedback literacies can be tied back to various steps of the assessment cycle.

Consider the direct relationship between self-assessment and peer-assessment. One limitation of engaging only with our own work is that we are so close to our own work that we can't see the flaws in it [235, 252]. If we skip steps, we don't even realize we skipped them because we fill in the gaps automatically. This also allows us to get away with less than coherent arguments, because we give ourselves the benefit of the doubt in ways we wouldn't with a peer. A similar logic applies to our writing. Having a second pair of eyes look for typos, grammar, or overall coherence is much quicker than reading and re-reading our writing ourselves. As we practice our skills of assessment with the work of others, it helps us build our overall practice of evaluative judgment [292], which we can apply to our own work.

Just as reading our own work isn't the best way to learn to self-assess, reading complete and correct work from an expert isn't that helpful. While it is good to have models of competent performance, what we really need is ample examples of work that is of varying quality [299]. When we read a solution from an expert (i.e., our teacher), we expect it to be correct. We don't expend much effort into deciding whether it is correct. Yet, when we see the work of a peer, we implicitly know that it might be lacking in one way or another, so we try to grapple with the work and decide for ourselves whether it is correct. If we want our students to learn to self-assess, a very effective avenue to learning to do so is through assessing the work of their peers.

6.3.2 Peer-Assisted Reflection

Here I introduce a structured routine for implementing peer feedback, called Peer-Assisted Reflection, or PAR [219, 235]. In an early study, I found that the use of PAR in introductory calculus increased the passing rate from 56% (the departmental average) to 79% [219]. I was also able to connect the PAR intervention to improvements in student explanations, with an effect size of ($d = 0.62$) [221]. Even though these studies were based on quasi-experiments with comparisons, they provide strong preliminary evidence in favor of PAR, consistent with meta-analyses of peer-assessment for learning.

I developed the PAR procedures over many years and found that it was helpful both for catalyzing learning and peer- and self-assessment skills. Since then, I have used PAR across a variety of contexts, from introductory calculus to proof-based geometry, and graduate analysis. Colleagues have used this technique in other settings, such as linear algebra [34]. PAR can be incorporated into students' independent problem-solving activities (e.g., through weekly homework assignments). Overall, this is a flexible strategy that can be adapted for use across settings.

If your students do not have prior experience with peer assessment, there will be an initial learning curve for them to start developing feedback literacies and for them to understand the overall logistics of the cycle. Because this is a new activity for many students, I have found that it is best to incorporate PAR as a weekly activity. This ensures that peer assessment becomes a regular part of your course, and that students have regular opportunities to build their skills. If you were to use PAR only a few times throughout the semester, the impact would be diminished because students would not have an opportunity to develop the requisite feedback literacies.

To support the peer feedback cycle, I provide students with a PAR packet to organize their work (see Appendix A and Appendix B for sample packets for problem solving and proof, respectively).[7] A PAR packet consists of a cover page (with the homework prompt), a place for an initial draft, a place for reflection, a place for peer feedback, and then additional space to write up a revised solution. As students go through the process, they are asked to complete the following: 1) Write up a draft solution, 2) reflect on their draft, 3) exchange peer feedback, and 4) revise and turn in their solution. When students turn in their assignment, it includes the feedback they received and how they revised.

Logistically, I choose one day of the week to be a "PAR day" (regardless of how many times a class meets, this is just about 15 minutes). Let's say PAR day is Tuesday. Each Tuesday, students would bring their draft solutions to class. I would have students find a random partner (or I would assign one to them), and the partners would spend about 10-15 minutes exchanging feedback. First, I would give the students about 5 minutes to silently read their peer's work, to ensure they engage with what the other student wrote. At this stage there is no talking, not even clarifying questions (because students are trying to make sense of the written inscriptions). Next, the students have a chance to provide feedback based on what they read. At this point, students might clarify anything that was unclear. This is a source of feedback for the students, as they learn more about which aspects of their writing was unclear to their peer. Students might directly discuss feedback, compare approaches, or even try to collaboratively solve part of the problem they were each stuck on. After this feedback exchange, students would have a chance to revise their work independently and turn in the problem during a later day in the week. If you use PAR on a weekly basis, the learning curve to understand this process takes only a few weeks and then the students can engage very effectively. I will walk you through each of the steps with an example.

6.3.2.1 Initial Draft

To begin, students make an initial attempt to solve the problem. PAR will be much more successful if you choose interesting problems that support meaningful collaboration (see Section 4.4.1). For example, you want to have a problem that everyone can at least get started on (low floor), but that few students could complete correctly on their first try (high ceiling). Problems with multiple solutions, solution paths, or that require explanation and justification are also productive because they more provide opportunities for two students to show up to the conversation with different approaches. This allows for them to compare, contrast, and connect their ideas in a way that leads to deeper conceptual understanding.

One problem I have used in introductory calculus is what I call the "hand area" problem [219]. This problem focuses on the approximation of the area inside a student's hand and is introduced before any formal concept of integration has been discussed. Here is the prompt:

> In this problem, you will trace the top of your hand (to create a function) and approximate the area of the picture that you create. Your main tasks are to devise a method for approximating the area and to show that your approximation is very close to the actual area.

1. Put your hand flat on the grid provided (with fingers touching, no gaps) and trace the shape of the outline of your hand. Make sure that the shape you trace is a function (if not, erase the parts of the shape that would make it not a function).
2. Devise a method to approximate the area of the region inside the curve you have traced. Explain your method in detail and explain why it should work. (Don't perform any calculations yet.)
3. Use the method you described above to approximate the area of the outline of your hand. (Show your work.)
4. Describe a method for estimating the error in your method of approximation. (Error is something you would like to make small! Thus, an estimate for the error means being able to say the error is less than some value.)
5. Calculate an estimate for the error for your method.
6. Explain (in principle) how you could improve your method to make your estimate as accurate as one could want (i.e., minimize the error). (You do not actually have to perform the calculations, just explain what you would do.)

[7] I have also uploaded a Latex template for a PAR packet here: https://github.com/reinholz/PAR

This problem is a good candidate for a PAR problem, because every student has a different hand, and therefore will bring something different to class. Also, the ambiguity in how to trace the hand, whether they are creating functions, what the graph looks like, and so forth, all provide areas for clarification and discussion. Students produce a variety of methods to solve this problem. Some are more sophisticated than others, but there isn't a right answer that students need to find. Moreover, in my experience, the questions about error in approximations are quite challenging for students in introductory calculus and provide room for nearly all students to go deeper in their thinking.

A typical student solution might involve a student drawing squares on the inside of their hand. Here is a sample of work from a student named Peter, describing the error calculations for this type of solution:

> Since my estimate will be lower than the actual value (since I didn't include that were part inside and part outside of my function) I will estimate my error by adding up the amount of space I left out to make up "whole cubes" (cubes does not mean units, it means each individual cube on the graph paper) and decide how many more units^2 I should have had. Error should be within 5% of actual value.

This student solution has a lot of productive ideas in it, like trying to estimate the area from the inside as a lower bound. However, the ideas about accuracy and calculating error are only partially formed and needed to be refined. The peer feedback process provides space for that revision.

6.3.2.2 Self-Reflection

After creating a draft, students are asked to reflect on their own initial draft. In a problem-solving oriented course (like calculus), I might provide some of the following reflection prompts:

1. Did you show all steps?
2. Did you explain why, not just what?
3. Did you avoid the use of ambiguous pronouns?
4. Did you define all variables?
5. Did you provide appropriate diagrams, figures, graphs, etc.?
6. Did you verify if your solution is reasonable?

These reflection prompts were generated from multiple iterations of seeing what students struggled with. The purpose of these reflection prompts is to remind students about things they should consider in crafting a solution. As students engage with such prompts over time, they start to internalize these as important parts of writing a good solution and begin to check themselves on them. These prompts also provide guidance for the types of feedback students provide to each other.

During the self-reflection, students can also explicitly write down things that they struggled with or areas of the problem that they want feedback with. For example, in a proof-based courses, I would have students write two-column proofs (not in the traditional sense), where the right-hand column could be used for self-annotations and providing feedback [235]. This right-hand column allows students to annotate important definitions, diagrams, or connections in their argument that usually wouldn't be written in a formal proof. One student described this process as follows [224, p. 72]:

> Proof has begun to feel personal—I feel as if I have some say in how I might structure certain problems or at least present my work. Additionally, the PAR process has completely changed the way I approach proofs and the way I teach. I feel as if the annotations column and its free form have given me a voice and choice in proof that didn't exist prior.

The presence of the annotation column connects to practices of Universal Design for Learning (see Section 4.3.2), which allow the students multiple forms of expression in addition to their formal written mathematics.

6.3.2.3 Peer Feedback

The next step is for students to exchange peer feedback. As described above, this would happen during a particular day of the week, and students would have a random partner. Students spend about five minutes to read the peer solution and write comments, and another 5-10 minutes to discuss. The feedback process is fluid and allows for lots of productive ideas to emerge. For example, in the case of the student shown above, Peter, when he entered the conference with another student Lance, it turned out that Lance was trying to overestimate the area of the hand [219]. These two solution approaches interacted productively, and it helped the students realize that they could bound the function from both sides (above and below) to get a better error bound.

6.3.2.4 Revision

After receiving feedback and having the peer conference, students have an opportunity to revise their work and turn it in during the next class session. This is a critical step that helps students incorporate feedback into their learning cycle rather than simply receiving it and forgetting about it without ever using it. Here is an excerpt from the Peter's revised solution:

> To estimate error, I will make boxes of units^2 for the rest of the area that was not covered by the 28 units^2 from #2. This will allow me to account for all of the area I missed in my first approximation, plus any area outside of the curve that my units^2 cover. Since this will be an over approximation, I know that the true area under the curve will be less than the area I calculate by error.
>
> I know that the actual value for the area under the curve is lower than 44.66 units^2 but higher than 28 units^2.

In the revised solution, Peter calculates an underestimation of the area of the curve in part #2, by only using boxes that fit entirely inside the hand. He then calculated an overestimation by adding the boxes that partially covered the hand. Although Peter still hadn't formalized a method to reduce the error (through a limiting process), the idea of bounding from below and above is a productive mathematical idea, which will provide much more utility than simply trying to guess how much error was there (as in the initial solution), with no ways to put bounds on the error.

Overall, students may revise their work based on a variety of inputs: 1) discussion with their peer, 2) discussion with other classmates after the PAR discussion took place, 3) discussion with the instructor after PAR, and 4) new insights developed from further processing time. One advantage of PAR is that it ensures that students begin working on the problem a few days before they need to turn it in, which helps keep it on their mind and new insights develop over time. After the solution is revised, the students turn in the full PAR packet.

6.3.3 Feedback Literacy

For PAR (or other assessment activities) to be most productive, you can create activities to help students develop feedback literacies [177]. Feedback literacies focus on students' abilities to provide feedback to peers and to effectively utilize feedback they receive. Students can further develop their feedback literacies when they are positioned as active agents (rather than only recipients) in the feedback process.

My favorite way to support this practice is by giving students small samples of written work and asking them to assess and provide feedback on them. One way I incorporate this into my class session would be as an opening problem that students start working on as they enter class. Often, I will take samples that students have written to a problem (from a previous semester) that my students are already working on. This gives them familiarity with the problem situation so it's easier for them to provide feedback. Then, we spend a short amount of time with students individually thinking about how to improve the work, and then we discuss as a class. By comparing the sample work to an external rubric or standard, students enhance their evaluative judgment skills. Incorporating such an activity throughout the semester is not only a way to develop feedback skills, but it also requires different types of thinking than just solving a problem straight out.

I called this activity the "darts" activity, like throwing darts at a dart board. Here's a sample darts activity for the hand area problem: Instructions: Classify the "bullseye" (correct explanation), "on the board" (a mostly correct idea that is communicated poorly or has a minor error), and "off the mark" (incorrect) solutions.

Prompt: Explain (in principle) how you could improve your method to make your estimate as accurate as one could want (i.e., minimize the error).

Sample 1: If I took a limit as the width of the rectangles approaches 0 (making the number of rectangles approach infinity), the difference in the area under the curve and the rectangles would approach 0.

Sample 2: You could use midpoints rather than endpoints and it will be more accurate because there will be less overlap.

Sample 3: If I had more rectangles there would be less overlap and the approximation would be better.

These samples all come from real student solutions. None of the explanations are perfect, but some are definitely better than others. After students spend a few minutes looking at the solutions and determining what makes them better or worse (and more or less correct), the class has a brief discussion of each of the solutions.

6.4 Authentic Summative Assessments

Assessment practices that aim to describe what students have learned are known as *summative assessments*. Although grades are a common form of summative assessment, and one that is required for many of us, I would argue that grades rarely capture student learning in a meaningful or authentic way. In fact, the impact of grading is often in direct opposition to our goals of disrupting hierarchy, because grades are a powerful mechanism for (re)producing hierarchies. Alternatively, if we view assessment as a tool to promote social justice [167], we should attend to how assessment practices can empower minoritized students and consider how assessment practices can upend existing hierarchies.

Unfortunately, standard assessment practices in mathematics are notoriously procedural, stressful, and problematic [291]. The stereotypical image is students sitting at desks in a lecture hall completing a high-stakes exam. Answers are either right or wrong, and students have a limited amount of time to perform. This inauthentic situation is created solely for the purpose of grading students, and it is something that students will never do again once they leave school. This testing situation is riddled with access issues [193], and it reproduces inequities by invoking stereotype threats that inhibit the performance of minoritized students [280]. Moreover, it is inauthentic, and students are asked to perform in ways that have little relation to how they will use mathematics in their future careers.

Drawing from the perspective of assessment for inclusion [192], we can view summative assessments as something that we do in collaboration with students, not something that we do to students. This paradigm shift is much more congruent with a notion of students developing interdependence and independence, rather than performing for their instructor. When we involve students as partners in the assessment process, we enhance their assessment literacies. Indeed, this collaboration requires students to answer fundamental questions about what it means to know something, and how to demonstrate what you know.

An inclusive assessment approach benefits from principles of choice associated with universal design (see Section 4.3.2) [130]. When we co-create assessment opportunities that provide multiple options to students and allow students to customize assessments to meet their own needs and goals,[8] we are creating an assessment environment that is responsive to the needs and strengths of our students. As students transition from being students and move on to their future careers, they will benefit from knowing what their strengths are and how to productively leverage them. Here are some concrete examples of assessment practices that can be inclusive and allow for student voice and choice.

6.4.1 Portfolios

Student portfolios are my favorite alternative to final exams. Portfolios offer students an opportunity to showcase what they learned throughout the semester. While students would turn in their completed portfolio at the end of the term, I typically include 1-2 checkpoints throughout the semester to make sure students are on the right track and that they aren't waiting until the very end of the semester to assemble their materials. When I have students create a portfolio,

[8]Often when I include summative assessments on my syllabus, I include a disclaimer to the effect of "The specifics of this assignment are flexible and can be modified so that they best support your personal learning goals and access needs. Please discuss your ideas for potential modifications with me and I'd be happy to listen."

typically they are putting together a set of artifacts in a 3-ring binder that they turn in to me. I tell students to think of the portfolio as an opportunity to showcase what they have learned to me. It is also something that they can use to look back on and celebrate everything that they have learned. Here is a sample portfolio assignment that I have used for a proof-based course.

> You will use a portfolio to demonstrate your learning in this class. This will give you an opportunity to showcase your learning over time. It is also a physical artifact that you can keep for the future, documenting your hard work in this course. In addition to filing all your coursework, you will complete several portfolio items. I recommend compiling your portfolio in a 2-inch 3-ring binder.
>
> Here is the table of contents for your portfolio.

- Portfolio Cover Letter
- Table of Contents
- Math Biography
- Portfolio Items
  - A piece of work that you are particularly proud of.
  - A proof that you found really interesting.
  - A proof that you found difficult, but eventually understood.
  - Your "most epic failure" on a proof.
  - Choose 2 constructive proofs to compare and contrast.
  - Choose 2 deductive proofs to compare and contrast.
  - Choose 2 additional proofs and explain why you chose them.
- In-Class Activities
- PAR Homework Assignments
- Student Essays written to reflect on readings from the book Incompleteness

> Portfolio Cover Letter. In 1-2 pages (double spaced), describe your experiences in the course. What did you learn? What was surprising, interesting, or challenging? What will you take with you as you move on to your next steps as a mathematician or mathematics educator? This is your introduction to your portfolio, so say anything else that you need to say to make the rest of your portfolio make sense to the reader.
>
> Table of Contents. Complete a table of contents that describes what will be found in your portfolio.
>
> Portfolio Items. You will include a total of 10 proofs/problems to highlight in your portfolio. These could be from your in-class assignments or PAR homework. For these 10 items, put them in their own section apart from the rest of your course assignments, so they are easy to identify.
>
> For each portfolio item, you will complete a short "portfolio card." You should provide at least one sentence describing what your portfolio item is, and at least one paragraph describing why you choose this particular item. You may choose to type up all of these "cards" together if it easier, or you can attach physical cards (i.e., an index card) to each of your portfolio items.
>
> Math Biography, In-Class Activities, PAR Homework, and Essays: Simply include all these completed class assignments as a part of your portfolio. Remember, you are expected to write up a clear and complete solution for all problems that we worked on in class.

This example can be customized to your local context. The portfolio gives students an opportunity to reflect on their learning, and it gives you a much more holistic look at a student's learning process. This is also an artifact that students can later go back to revisit their experiences in the course. Portfolios are much more humanizing than high-stakes exams, and reading portfolios is far more enjoyable than grading exams.

While personally I have primarily used paper-based portfolios, electronic portfolios open a whole new set of opportunities for students to share and celebrate their learning. Students can integrate video, voice, music, images, and other modalities to fully express their learning and their humanity. Electronic portfolios also make it easier to create check points and provide ongoing feedback to students, without collecting a large stack of papers. Finally, the electronic formats allow both you and your students to keep a copy of the portfolio.

6.4.2 Projects

Projects aim to be an authentic representation of mathematical practice. Projects are longer term, substantive engagements that offer students more agency in the learning process. Projects allow for multiple opportunities for students to check in and receive feedback on their in-progress work. When students are provided with options—ranging from what their project will focus on, the timeline for their work, and how they will express their learning outcomes—it fosters meaningful partnership between students and the instructor. Projects are often easiest to conceive in statistics or applied mathematics, where there are numerous opportunities to analyze real datasets and model real-world phenomena. But projects in pure mathematics are also possible to create, and with some creativity, projects can be made accessible even to high school students [160].

Projects create opportunities for both individual and group-based engagement for students. Collaborative projects can help push back against the notion that mathematical performance is always individual and objectively quantifiable. As should be clear by now, effective group engagement can be bolstered with effective scaffolding, so thinking through group roles, responsibilities, community agreements, check ins, assessment practices, and so forth, should all be part of an explicit conversation to support groups to work on a project.

Industry and community-based partnerships also open up new opportunities for authentic projects [298]. For example, if a community-based organization has collected data but needs help analyzing the data and making sense of it, mathematics students can bring their quantitative skills to this problem in a way that both supports their learning and contributes to social good. This authentic context also requires students to consolidate their findings into a technical report or presentation that can be understood by the stakeholders, who do not have the same level of mathematical proficiency. As students customize their projects based on the partner institutions, they can engage in work that is personally relevant and meaningful.

6.4.3 Students as Teachers

There is an adage that the best way to learn something is to teach it to others. The learning benefits of teaching have been studied extensively, particularly in the literature on peer tutoring [82]. Similarly, offering students opportunities to teach their peers supports learning and provides an opportunity to assess student knowledge—both for you as an instructor, and for students as they reflect on their own understanding. I've created opportunities for students to act as teachers in my classes, in a variety of courses ranging from the history of mathematics to research in undergraduate mathematics education. A similar idea can be used in math content courses too. Students take time to independently research an idea and create a learning opportunity for other students to engage more deeply. This allows students to draw on a variety of creative skills in lesson planning and can build interdependence through group planning. I typically position students as instructors, rather than ask them to give a formal presentation, for two reasons. First, this positions students as experts, rather than putting them under the pressure of an evaluation. Second, this provides students with more flexibility to focus on engaging students in the class, rather than just lecturing.

Many variations on this theme are possible. Students could prepare a formal talk, like one they would give at a conference. This might be useful if presentation skills are a learning goal of yours. Preparing and delivering such a presentation also requires students to do the math, consolidate their thinking, and make it visible to others. Of course, presenting publicly can be stressful for many students, so you'll need to create a safe environment, providing scaffolding and support to build the required skills. Another alternative is to have students create and record a digital presentation. Or you can have a poster session where many students are presenting their ideas and interacting with one another. All these options allow for flexibility in how students share their knowledge, and position students as competent contributors to your class.

6.4.4 Multimodal Essays

Essays are a valuable way to help students further develop their epistemology of mathematics and explore the historical development of mathematics in more depth. If you are humanizing mathematics by including stories and biographies of real mathematicians—historical and contemporary—essays provide one mechanism for reflection on those stories. While essays are not typically seen as an assessment that would be used in mathematics, I would argue that essays provide a valuable opportunity for students to improve their technical writing and communication skills. Because technical communication is so important to success in mathematically intensive fields (e.g., working in industry), providing opportunities for students to practice has great value. Student engagement can be further supported by clear assessment criteria, through rubrics, learning goals, and class discussions.

Essays traditionally consist of a written product, but there are also creative opportunities to utilize digital media. For instance, rather than creating an essay, a student could create a website, blog post, or even Wikipedia article. Such opportunities allow students to share their thinking with a broader audience and may even allow for forms of public engagement. Students might also use mixed media that consists of text, images, video, voice, and so forth. In an increasingly global and digital world, building these digital literacies is another valuable form of learning for students, which can be integrated with their mathematical development.

6.4.5 Using Rubrics

Authentic assessments can be enhanced rubrics. Rubrics help students understand in more clarity what the learning goals they are trying to reach are. For example, Johanna Rämö has incorporated rubrics into the teaching of linear algebra to help students better understand the learning outcomes for the course [217]. This rubric lays out some of the major skills of the course (e.g., vector spaces, matrices, geometry, MATLAB, reading and writing, discussion, peer feedback) and the continuum for what student skills would look like from emerging to higher levels. This helps students gauge where they are, where they need to go, and how to get there. It also allows for the targeted design of assessments that align with these particular learning goals, so that students can get more targeted feedback about what they do and do not understand [154, 215].

Instructors can also directly involve students in the process of generating rubrics [7, 127, 131, 179]. Although much of this work has happened in writing or in the social sciences, these techniques can also be applied to mathematics. For example, when my students are expected to produce portfolios, we also have a class discussion about what would make a good portfolio, and we construct an assessment rubric together. Just like how the PAR activities help students build a general sense of solution quality and self-assessment, having students involved in the process of rubric generation—either for problem solving or larger projects—gives them agency in the process and supports assessment literacy.

6.4.6 Exams

Ideally, I would not use exams at all. However, due to institutional policies, pressures, or other reasons, you may be required to use exams, or you may choose to use them. Here are some suggestions for making your use of exams more equitable. Timed exams put a lot of pressure and invoke stereotype threat. A take-home exam provides an alternative with less time pressure. Allowing students to use their resources (notes, etc.) is another way to take off pressure and create access. Group exams create another opportunity for students to build interdependence as they work together on the exam. Exam corrections are another strategy to reduce pressure.

6.5 Alternative Grading Approaches

Although I would ideally avoid grading altogether, practically, most institutions require us to assign grades to our students. As described in Section 6.2.4, when grades are combined with written feedback, it reduces the value of that feedback [135]. A corollary to this is that we should provide many opportunities for assessment and feedback in our classes that aren't tied to grades. Then, we can carefully choose what to grade in our class and how.

6.5.1 Why Grading is Problematic

Grades reduce student learning and motivation [135]. Moreover, even though grades are the primary currency of the testing industrial complex, there is no standard meaning for grades. Grades typically measure some combination of the following factors: 1) performance, 2) learning, 3) effort, 4) attendance, 5) participation, 6) accumulated privilege, 7) the ability to "do school," 8) the ability to "test well," 9) a comparison to peers, and 10) random noise. If you were to look at a students' transcripts and see their prior grades, how do you interpret them? You have no idea which ingredients were put together to construct the grade. The most you can infer is that the student was previously successful in school, which may predict that they will do well in school later, but it doesn't necessarily mean that they have a deep understanding of mathematics. Especially when we consider standardized assessments, we need to consider a variety of racial [88] and gender [57] biases, as well as culturally-based language demands [1].

In addition to being a bad measure of learning, grades (and testing more broadly) have a variety of negative consequences. Testing regimes narrow the curriculum, put pressure on teachers to cover content, and overall, stifle joy, creativity, and innovation in education. A typical calculus course covers so much content that topics are covered at a surface level and students can barely keep up with it.[9] Because calculus is viewed as a service course, instructors are required to test students on a wide array of topics to prove that they are ready to move onto appropriate major courses. The combination of testing regimes, historical precedence, canonical textbooks, and institutional pressures with tightly coordinated curriculum allow very little room for change, even when most people can agree that the status quo isn't working for most students.

It is also difficult to create valid and reliable measures of learning in calculus, or in any content area, for that matter. Consider what it means to understand a limit in calculus. *Does this mean someone can compute limits? Or describe limits conceptually? Or answer questions about how a limit applies? How many questions does someone need to answer correctly? Are they procedural, conceptual, or both? Does someone need to prove theorems using limits? How complex do those theorems have to be? Is understanding a limit tantamount to working with an epsilon-delta definition, or does that not matter in calculus? What is the difference between understanding limits in calculus as compared to real analysis?* Clearly, measuring understanding of limits in calculus is nontrivial. We often assess students by providing a carefully chosen sample of limit problems, but there's little theoretical or empirical basis for those questions as being a good measure of learning. Even students who ace calculus exams may be described by the professors of later courses as not understanding calculus at all.

Suppose that we do come up with some agreed upon measure to assess understanding. Can we reliably implement that measure across students and all variations of understanding that they display? And if we have multiple assessors, can we get them to agree (i.e., do we have interrater reliability)? Even under ideal circumstances, it's difficult to accurately measure learning. As a fact of life, we must accept that as much as we would like to, we cannot really measure student learning as accurately as we would wish.

One way to move forward is to divest from the notion that all learning is individualized and quantifiable. Instead, we can focus on collective progress, or what our students can do collaboratively with other peers and with appropriate support [194]. As we take up these alternative approaches to our assessments, it allows us to consider how assessments can be tools for student learning, social justice, and broader inclusion. Assessment doesn't need to be seen as something that happens after learning, but it is a very powerful tool to produce learning.[10] Assessment isn't just for a grade, but rather, it can be seen just as a part of the learning cycle.

6.5.2 Standards-Based Grading

Standards-based grading focuses on measuring student progress on a set of well-defined learning objectives [121].[11] Often, this involves generating a set of rubrics that shows progress along a particular number of dimensions. Although it is not broadly used, standards-based grading is probably the most widely used alternative grading system in mathematics education. For example, there is an entire special issue in *PRIMUS* describing how instructors have adopted

[9] In fact, many students who take calculus in college have already taken it in high school, which allows them to keep up [27].

[10] For example, consider that practice testing in a low stakes environment is one of the most effective ways of studying.

[11] Standards-Based Grading is sometimes called "Mastery Grading," although this description is increasingly being phased out given the etymology of the term Mastery and its roots in enslavement.

this type of approach in postsecondary mathematics [35].[12] Rather than averaging scores on a variety of assignments (primarily exams), standards-based grading requires that students complete work demonstrating a particular standard along a variety of topics [208]. One of the benefits of this approach is that it lowers the stakes of any individual assessment, because students can continue to improve their quality of work along any given standard throughout the semester. This means that students who are coming into the class with less understanding of the material aren't punished for their initial understandings but are graded more based on where they end up at the end of the semester. Although students still may have more work to do to reach the standards, they also have more opportunities during the semester to make up for their lack of initial understanding.

There are a wide variety of ways to implement standards-based grading, depending on how final scores are computed. For example, instructors might choose which types of assignments and which types of problems can contribute to scores that meet the standards. Similarly, instructors need to decide to what extent students can revise coursework, or if they simply take new assessments on the same topic. In the ideal, standards-based grading would allow for unlimited attempts to reach the highest levels of competency, but given logistical constraints there is typically a limit to what is feasible for a single instructor to do in a single semester.

6.5.3 Specifications Grading

Specifications grading is similar to standards-based grading, but it differs insofar that a simple, common rubric is used for most assessments [37, 196]. One example of such a rubric is the EMRN rubric [46, 294].[13] The EMRN rubric is essentially an elaborated version of a Pass / Fail rubric with four categories:

E: Excellent/Exemplary: the work exceeds expectations and could be a model for others

M: Meets Expectations: the work is mostly correct and communicated reasonably well, although it could be improved with revisions

R: Revision Needed: there is partial understanding with significant gaps remaining

N: Not Assessable: too much work is missing to assess understanding.

To make EMRN grading (or any other specifications grading) effective, an instructor needs to have a way to communicate the expectations to students. This is often done with a rubric of sorts that describes high-quality work. This could also be achieved with exemplars that highlight different levels of understanding. With an approach like this, an instructor can quickly grade student work in essentially a pass/fail manner. Like with standards-based grading, a typical way to implement specifications grading would be to incorporate opportunities for students to revise their work if it is deemed revision needed.

A general principle for using specifications grading and managing workload is to consider each assessment revision like a new assessment (Spencer Bagley, Personal Communication, June 2022). It follows that if you plan to have students revising their work in your class, you will want to reduce the number of assessments you use overall. The EMRN rubric can also be simplified to essentially M/R, where something either meets expectations or needs to be revised. To compute a final course grade, these specifications-based assignments can be combined into bundles (i.e., different learning targets) to generate actual grades. An example of how to do this, and further information, can be found on Robert Talbert's blog [293].

6.5.4 Ungrading

Grades can negatively impact student identities (e.g., Receiving a D reifies the identity of "bad at math" for a student), they can be biased by stereotype threats, they undermine intrinsic motivation, and overall, they reinforce an individualistic, capitalist environment in which students are forced to compete with one another rather than work together. Given all of these issues, how do we assign grades in a more just an humanizing way? One mechanism is to use *ungrading*, in which students and the instructor are collaborators in determining student grades [22]. In practice, this might involve

[12]There is also a useful set of web resources hosted by College Bridge: https://college-bridge.org/our-services/conferences/the-grading-conferences/resources/

[13]This is a revision on the EMRF rubric, where the F was renamed to remove the association with failing.

students writing multiple reflective memos throughout the semester in which they describe what progress they feel that they have made. This progress is enhanced when there are explicit learning criteria, for example in a rubric, especially if that is co-constructed between students and the instructor. The process of reflective writing involves students actively in thinking about what they have learned and how that translates to a grade. Spencer Bagley [8] has a public syllabus in which he shares his policies for using ungrading in teaching the Calculus of Variations,

> Final letter grade.
>
> You may have noticed that in the welcome survey, I asked you what you think should characterize the work and participation of an A student, a B student, and a C student. This is because at the end of the term, you're going to tell me how you did and thus what grade you earned in the course, based on the following rubric we're going to create together.
>
> **Description of work and participation for grade.**
>
> A Show up to class if people are expecting you. Put forth work and effort. Participation: Speaking up in class or in groups or on discussion boards or whatever. Be a helpful part of the discussion. Be respectful and provide a good learning environment. Complete assignments and quizzes on time. Revise assignments when necessary; do really good reflections. Prove mastery of the course material. (Maybe not everything; that's ok! Needs to be an achievable level.)
>
> B Participates majority of the time. Completes most assignments and quizzes on time. Understanding mostly at mastery level, but their work maybe doesn't reflect their understanding clearly enough to rise to A level. Maybe a good chunk of work that needs to be revised but isn't.
>
> C Similar stuff as A, but only like half the time. Inconsistent: maybe bursts of effort that aren't really strung together. Participation: Lackluster. Often absent, or doesn't speak up when present. Still maintains a respectful classroom environment. Doesn't consistently revise assignments that need to be revised. Superficial understanding of course material. Overall understanding at "revision" level.
>
> D Let's not be here, but less than half the work of the course.
>
> F (Let's really not be here.)
>
> You can use all your assessments throughout the course as evidence in this discussion. I will probably agree with you, because you know yourself, your work, and your understanding pretty well. If I don't, we'll talk about it.

As the course policy states, determining the final grade happens through a conversation between the instructor and the student, in which the student has an opportunity to demonstrate what they have learned. This type of approach shifts typical power dynamics and allows students to play a more active role in the grading process.

6.5.5 Putting it All Together

Alternative assessment and grading practices aim to upend the hegemony of high-stakes exams. There's no single way to determine grades, and these strategies can be used together. A single course could use standards-based grading with rubrics, and some of those standards might be elaborated rubrics, while others are simply meets expectations/revise. How grades are determined could be in conversation with the students, or it could be simply based on computing whether students met a certain number of pre-determined objectives to the required level. Learning to use alternative grading practices will take time and probably won't work perfectly the first (or second) time that you do it. Although these methods are imperfect, they are still preferable to traditional grading schemes, which are known to be flawed for myriad reasons. If you build in a lot of flexibility, revision, and room for growth in the grading process, you can better leverage assessment to support the learning process.

6.6 Reflection

Now that you have reached the end of this chapter, here are some questions and activities to support your further reflection. These questions will be most effective if you consider them in collaboration with others, who will have their own and different responses.

1. Think back to the guiding question of the chapter: Based on the evidence in this chapter about what types of feedback support learning, are you willing to change your course policies and grading policies?
2. Your partner (or a close friend) has just cooked a meal for you but it tastes very bland because they forgot to put any salt in it. Write a dialogue about how you could provide this feedback in a way that is supportive and does not offend them.
3. Think of a situation where you had to give or receive difficult feedback? How did that go? Based on this chapter, what ideas could you use to make that situation better?
4. Think about the last time you went through a peer review process. In what ways did the feedback provided to you meet or not meet criteria for productive feedback for learning?
5. If resources were not a limitation, what would be the most ideal assessment you would give your students to see what they know? Have fun and be creative.
6. Think about something that you know really well. How do you know that you know it? How could you prove it to others that you know it?
7. When was the last time you praised somebody? If you could go back and change that interaction, how would you change it?
8. In an ideal world, what would you do instead of assigning grades? Can you take any of those ideas into your own teaching practice?
9. Looking at the references of this chapter, what is one reading that is new to you that you would like to read in its entirety?

7

Catalyzing your Learning

Up to this point in the book, I have focused on a variety of different teaching practices that you can use to disrupt hierarchies in your classroom. Consistent with the simplified learning cycle, this would constitute the instruction portion of your own growth as an instructor. As you know, simply receiving that instruction will do little to nothing to change your teaching in a lasting way. In fact, you could read through the earlier portions of the book in a day and feel great about everything you "learned," but unless you put in the weeks, months, and years to become proficient with these techniques and integrate them into your repertoire of practice, this book would do little more than give you a false impression of learning. Let's go deeper. Here is some practical advice on how to use this book as a resource for you to practice and improve your teaching techniques and receive feedback on your practice.

Before you read this chapter, I'd like you to take a minute to write down some ideas that you have read in this book. What are some of the key concepts and practices that you've encountered that you'd like to incorporate into your own teaching? Thinking about those practices, which are the ones that you could implement tomorrow, and which do you think you would need more support to do effectively? Can you think of any colleagues or other support structures that could facilitate your learning? Please take out a piece of paper (or use an electronic device) and take five minutes to write these things down. This extra processing time will make it easier for you to see what is new to you. In addition, this creates an opportunity for you to write down *new ideas* in this chapter as you encounter them. As you read through this chapter, I want you to reflect on the following question: *What is a concrete step you plan to take to be in community with your colleagues and work together to improve your teaching?*

7.1 How to Practice

A first step to improving your teaching is to set reasonable expectations for yourself. When I work with faculty in a professional development context, I take an incremental approach to improvement. In an EQUIP learning circle (see Section 7.3.2), I perform monthly classroom observations for each participant. Between each observation, I provide data and concrete suggestions to the participants. My goal is for them to pick *one* new teaching practice that they would like to work on before our next meeting (although some do choose more). Practically, this means that participants spend a few weeks building fluency with a single practice at a time. I have found that this is much more effective than trying to change everything at once. When you try to do too many things at once, you're unlikely to do any of them particularly well. Also, from a scientific standpoint, changing too many variables makes it difficult to track the impact of any individual change. Using any new idea takes time and practice!

Even though the faculty members I work with may only change 3-4 practices over an entire semester, the impact on their students is profound and measurable. Just by making a few small but intentional changes to your teaching, you can have a large impact on disrupting hierarchies in your classroom. If you continue at this pace for multiple semesters (and even years), you will slowly but surely transform your teaching in powerful ways over time. Not only are you promoting social justice, but teaching also becomes easier and more enjoyable. You will need to spend less time preparing, and your students will appreciate your classes more. It takes time and effort, but the benefits will last your whole career.

You may be able to read this book in a few days, but learning the lessons it teaches will take years. It's just like reading a mathematical text, in which you might spend five hours decoding the mathematics on a single page. In this case, there is less decoding, but a whole lot more teaching practice and revision on practice that must take place.

7.2 How to Get Feedback

You can draw upon multiple sources of information to get feedback on your instruction: yourself, your students, student assistants, faculty peers, instructional coaches, etc. To begin disrupting hierarchies in your classroom, you need to know what the hierarchies are. Although my empirical work across hundreds of classrooms shows the ways in which non-disabled White men (and some Asian students) dominate discussions in US math classrooms, the ways in which power and privilege play out in your classroom will be unique to your local context. Thus, while having a general understanding of the types of hierarchies that are likely to play out is useful, to truly make these teaching practices effective, you need to have local data. In my own work, I have used measures of participation (through the EQUIP tool) as a measure to provide such data, but this is certainly not the only such measure.

You may be tempted to rely on your gut impressions of who participates and not. I strongly advise against this. Our self-perceptions are mired with implicit bias and when you are teaching, you are simply managing too many things to develop an accurate perception of who is participating and how. Instead, you need someone to observe your classroom and actually tabulate student participation. If you have a student assistant or faculty peer, they could sit in your classroom and create a map of your students to track participation. Another method is to record your own teaching and analyze it yourself. If you've never recorded yourself to watch yourself, I highly encourage it. This is one of the most useful things you can do as a teacher to get a better sense of what you look like in the classroom. If you are going to record yourself, set up a video camera in a way that it captures your students (not yourself), so that you can see exactly who participates and how. Before you do this, inform your students that this is for your own learning and not to be shared with anyone else. If certain students object to the recording, you can move those one or two students out of camera view.

As a first step to understanding hierarchies in your classroom, you can simply tabulate student contributions (see Figure 7.1) and even better, how they are distributed amongst student identities (e.g., you can start with your gut impressions of racial and gender identities, or you could send out a survey to your students to ask for demographic information). An observer can also annotate the drawing with student names and other things as they come out during the discussion. To go deeper, you could use a formal observation protocol, such as EQUIP. EQUIP Provides a free, customizable web app that will automatically generate data visualizations for you. EQUIP allows you to go much deeper than raw counts to understand the quality of participation and how that is distributed. It also has features to help you track your improvement over time.

These quantitative data about participation patterns are one of the primary ways you can get feedback about your teaching. My whole methodology of instructional change is organized around these data. When you identify the hierarchies and inequities in your classroom early in the semester, it will help you identify exactly the leverage points you need to start implementing the practices that we discussed earlier in the book. Thus, rather than implementing teaching strategies in a general sense, you can do so with social marker specificity. For example, if you have identified a specific racial inequity in your classroom, you would use strategies to promote greater participation from that group.

You can also gather useful data from your students. For example, Section 4.1.3 describes strategies you can use for checking in with your students to get feedback. Similarly, Section 6.1 describes formative assessment strategies that you can use to elicit student understandings and feedback. It is helpful to approach student feedback with humility, acknowledging that our work to improve as teachers is always ongoing, and that you value the perspectives of the students. When students do give you feedback, you can be responsive to it by explicitly sharing how you plan to change, and then implementing those changes. This will encourage students to give more feedback.

The last source of feedback would come from a peer observation of another faculty member or instructional coach. Peer observations will be most useful if you have a clear learning objective or goal (just like when we give feedback to our students). Thus, a first step is to determine the skills or objective for the peer observation. You don't want to ask,

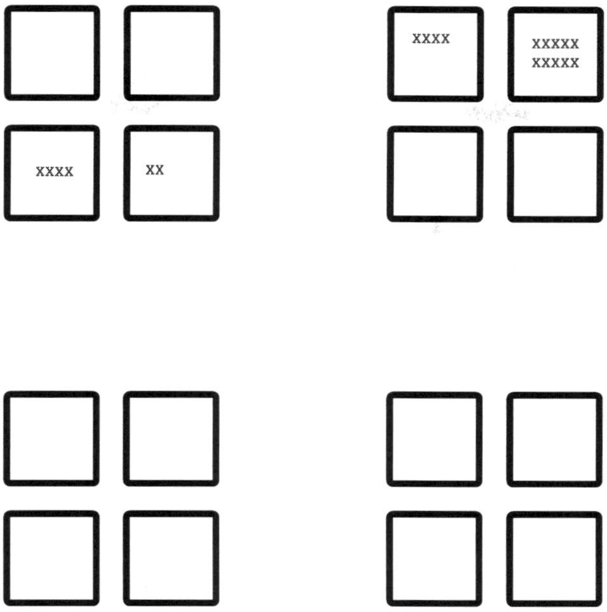

Figure 7.1. A sample paper and pencil observation map. Each box is a single student, and each X represents a student contribution.

"am I teaching well?" but something like "when I ask students to give an answer, which students raise their hands?" or "When students offer ideas to the class, how am I using student thinking?" Such questions focus on concrete practices, rather than subjective perceptions. The observer could also use a formal protocol such as EQUIP to support your learning.

7.3 Learning Communities

Learning is a social process that can be greatly enhanced by working with your peers. Although you can still benefit from this book as an individual, that learning can be enhanced when you do it with your colleagues.

7.3.1 Book Club

The simplest approach you can take would be to start a book club. This requires very little organization, and it creates a dedicated space for you to discuss ideas contained in this book. It will also help you build community with your colleagues and have a place to talk about issues related to teaching, which we often do not have on our campuses.[1] In addition, if you're aiming to start up a book club, it is likely that you can get a small amount of funding from your department chair, Center for Teaching and Learning, or other unit on campus that supports teaching. I suggest meeting roughly once a month as a learning community. Meeting for an hour at a time will still allow for enough time to get into the substance of the work, without feeling like too much of a burden. You might even schedule the book club meetings over coffee or lunch as a way to humanize the process and build further community.

The book club meetings will provide checkpoints for the practices you're trying in the book. Between meetings, I suggest that you choose a single teaching practice that you'd like to work on, that you can talk about with your colleagues the next time you met. This will allow you to ground the ideas shared in this book in concrete classroom experiences. Another idea is for your entire group to agree on a single teaching practice that you'd like to work on, so you can all share your experiences and support one another to improve. To encapsulate your learning (and to secure funding from relevant administrators), consider offering to present your learning at an appropriate venue on campus (e.g., a faculty meeting, department colloquium, seminar for a Center for Teaching and Learning).

[1] You may even consider opportunities to get students involved in the reading circle as well.

7.3.2 EQUIP Learning Circle

An EQUIP learning circle is a structured professional development approach that I have used for many years now. The frequency of meetings is similar to a book club: roughly each month for an hour at a time. However, in a learning circle, there is also a coach or facilitator who plays a role in facilitating the conversation and organizing data collection and analysis around patterns of participation, using the EQUIP tool. Often, the facilitator would have a student assistant to help collect and analyze data, and provide feedback. Ideally, the coach could come from a campus unit such as a Center for Teaching and Learning. Other options include having a faculty member serve as a peer mentor. This is especially effective if someone has already been through the EQUIP process previously as a participant. Other options are to have faculty members code each other in pairs, or even code their own videos to generate data. Regardless of how the data are generated, this becomes a collective processing space for making sense of hierarchies that the data reveal and developing steps to address them.

7.3.3 Collective Action

Change doesn't happen in a vacuum. What we do in our classroom is important, but students are still affected by the world around them (including the rest of our campus community). To make lasting changes, we are much more effective when working in community. Learning communities can offer an opportunity to build capacity and political power that extends beyond any individual. For example, if your community comes up with concrete suggestions to improve your work, you have a stronger basis on which to ask an administrator for funding. Given that improving equity in STEM has become a national imperative in the US, there are many options to tie into existing resources in order to support your efforts. For example, *Can the institution provide stipends? Or can it provide student support to help with coding? Or special training for participants and others?* All of these concrete asks could support your work of collective disrupting hierarchies.

7.4 Putting it All Together

This book provides a wealth of practical strategies you can use to disrupt classroom hierarchies. This process begins with awareness. First, you need to have awareness of yourself, your positionality, and how you show up in the classroom. Second, you want to educate yourself about the types of hierarchies that commonly arise in classrooms. Third, you should collect local data to help you understand how hierarchies emerge in your classrooms and how that changes over time. Finally, you can start implementing teaching practices in this book and track their impact over time.

As you set up your classroom environment, you'll want to do so with intention and build trusting relationships with your students that support authenticity, vulnerability, and risk taking. You're creating a safe and accessible environment that acknowledges some students have had very negative mathematics experiences, or even trauma. It can help to be explicit about the the racist, sexist, and ableist status quo, and describing your commitment to doing things differently. Beyond empty talk, you need concrete actions that you can implement.

As you create that classroom environment, you can use a lot of teacher moves to position your students differently and elevate their status. Again, this starts with knowledge of the students, who is participating, and whose ideas you need to elevate. This social marker specificity allows you to use strategies to intentionally impact certain groups of students (e.g., by race, gender, disability), to have a greater impact. One of your most powerful tools will be asking questions. Changing your questioning strategies takes time and practice, so you might write down good questions you plan to use (and who you plan to ask) in your lesson plans. Your goal is to move away from simply seeking answers (what questions) to dig deeper into the depths of student thinking. Student ideas should form the basis of your discussions, and the only way that can happen is if you get student ideas out on the table. Using strategies like wait time, five hands, or think-pair-share, we can get past the first and loudest students dominating our classrooms.

After you get student thinking on the table, you must figure out what to do with it. The way you respond to what students will shape the course of your discussion. And over the longer term, it will shape which students are seen as competent, and who willingly volunteers to participate or not. One of your greatest assets will be to look out for the moments in which interesting ideas come out so that you can draw attention to the competence of specific students who may not participate often. Rather than spontaneously starting discussions and hoping for the best, an even more effective strategy is to orchestrate the discussion before it happens. This allows you to craft a beautiful and productive discussion that will seem to be mostly driven by the students, but really you are using your role as facilitator to get the right student ideas out at the right time. When you craft a discussion in this way, you have the ultimate control of who participates and how, which means you can give students opportunities to succeed that they otherwise might not take up on their own.

Building on this classroom participation, you'll want to have mechanisms for assessing students and providing meaningful feedback. Engage your students as partners in the assessment process so that they can have greater authority and develop as independent learners. Focus on the types of feedback you give, emphasize growth and competence, while moving away from empty praise. Be sure to decouple the grading and feedback processes, as grades undermine the value of feedback. When it comes to summative assessments and grading, look for more humanizing alternatives to support your students.

This process will take time and practice. It can be frustrating at times. Your teaching won't change overnight, but as you start learning to implement new practices, you can have a big impact even with small changes. You don't have to do this alone. Leverage your peers and community resources in a way that can support you and build collective action for disrupting hierarchies. You are not alone, and your students will appreciate the effort you put in to teach in a more equitable and humanizing way.

7.5 Reflection

Now that you have reached the end of this chapter, here are some questions and activities to support your further reflection. These questions will be most effective if you consider them in collaboration with others, who will have their own and different responses.

1. Think back to the guiding question of the chapter: What is a concrete step you plan to take to be in community with your colleagues and work together to improve your teaching?
2. Record yourself teaching and watch the video. What do you notice?
3. Identify a colleague in your network who can perform a peer observation for you.
4. Invite a colleague out for coffee to share with them some ideas you have learned from this book.
5. Identify a sympathetic administrator who might be willing to support your efforts. Have a meeting with them to share the work you're doing and your vision for transforming teaching. Can you get their support?
6. Find a group of colleagues to run a book club with. Run the club!
7. Form an EQUIP Learning Community with your peers. If you need help getting started, I would love to hear from you (daniel@danielreinholz.com).
8. Think about the last time you learned something in community with others? What was it, and how did that experience go?

A

Sample PAR Packet (Problem Solving)

Name: ________________________ Section: ________________________

The PAR Process

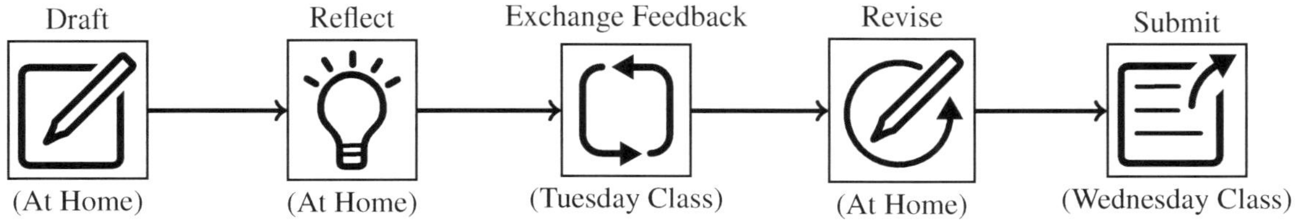

Problem Statement (Filling Bottles)

You are filling up bottles with liquid coming from a tap at a constant rate.

1. For each bottle, sketch a graph of the height of liquid in the bottle as a function of time.
2. For each bottle, sketch a graph of the rate of change of the height of liquid as a function of time.

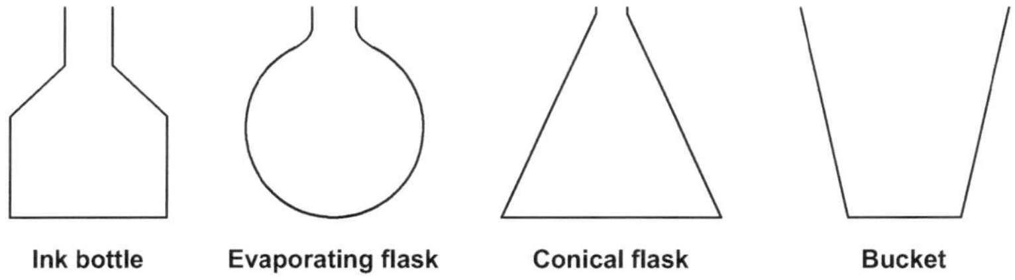

Reflection

Turn the page and check off the icons for things you think you did well; circle the icons for things you would like feedback on.

Feedback Provided By: ____________________

Suggestions	Communication	Strengths

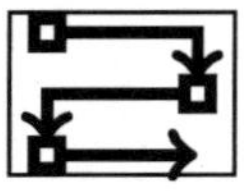

Show All Steps

Explain Why, Not Just What

Avoid Pronouns

Use Correct Definitions

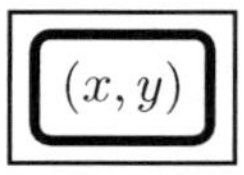

Define Variables, Units, etc.

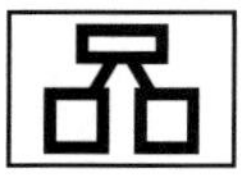

Create Diagrams

Suggestions	Accuracy	Strengths

Correct Setup

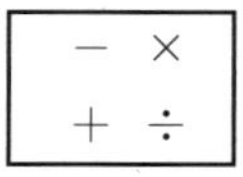

Accurate Calculations

Solve Multiple Ways

Answer Reasonable

Other (Write Below)

Draft (Page 1) Name: ____________________

Draft (Page 2)

Submission (Page 1) Name: ______________________

Submission (Page 2)

B

Sample PAR Packet (Proof)

Name: ______________________ Section: ______________________

The PAR Process

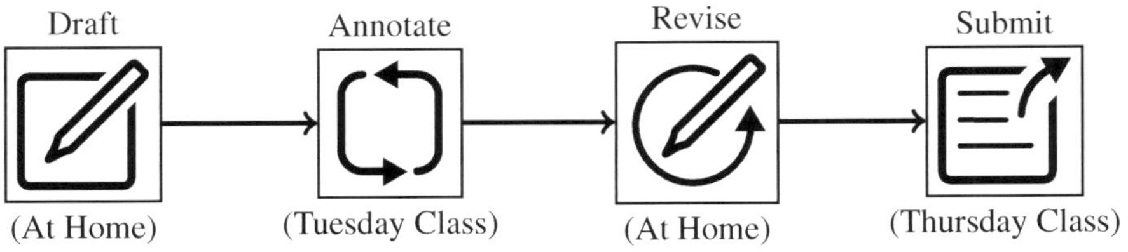

Write your solution in the left column. The right column is used for annotations. If you provide feedback to your peer, you will annotate their solution. After class, you will annotate your own solution as well (and write "self" on the annotated by line). In your submission, use the annotation column to explain how you did (or didn't) response to peer feedback.

Problem Statement: Continuity is a point-wise property of functions. In class, we also defined *uniform continuity* as a property across the entire domain of the function.

1. In your own words, describe the difference between continuity and uniform continuity.
2. Consider the function $f(x) = x^2$. On what domains is it continuous? Prove your conjectures using the definition of continuity and its negation.
3. Again consider $f(x) = x^2$. On what domains is it uniformly continuous? Prove your conjectures using the definition of uniform continuity and its negation.

DRAFT

Solution **Annotated by:** ______________________

SUBMISSION

Solution **Annotation**

C

Bibliography

[1] Jamal Abedi. Standardized achievement tests and english language learners: Psychometrics issues. *Educational assessment*, 8(3):231–257, 2002.

[2] Olusola O. Adesope, Dominic A. Trevisan, and Narayankripa Sundararajan. Rethinking the use of tests: A meta-analysis of practice testing. *Review of Educational Research*, 87(3):659–701, 6 2017. publisher: American Educational Research Association.

[3] Subini Annamma, Deb Morrison, and Darrell Jackson. Disproportionality fills in the gaps: Connections between achievement, discipline and special education in the school-to-prison pipeline. *Berkeley Review of Education*, 5(1), 2014. [Online; accessed 2021-02-11].

[4] Alfredo J. Artiles. Fourteenth annual brown lecture in education research: Reenvisioning equity research: Disability identification disparities as a case in point. *Educational Researcher*, 48(6):325–335, 8 2019. publisher: American Educational Research Association.

[5] Wayne Au. *Unequal by design: High-stakes testing and the standardization of inequality*. Routledge, New York, NY, 2010. [Online; accessed 2016-11-10].

[6] Wayne Au. Hiding behind high-stakes testing: Meritocracy, objectivity and inequality in us education. *International Education Journal: Comparative Perspectives*, 12(2):7–19, 2014.

[7] Ruth Bacchus, Emma Colvin, Elizabeth Bronwen Knight, and Leonora Ritter. When rubrics aren't enough: Exploring exemplars and student rubric co-construction. *Journal of Curriculum and Pedagogy*, 17(1):48–61, 1 2020.

[8] S. Bagley. Math 300c syllabus may 2020, 2020. [Online; accessed 2022-06-13].

[9] Rosangela Bando, Emma Näslund-Hadley, and Paul Gertler. Effect of inquiry and problem based pedagogy on learning: Evidence from 10 field experiments in four countries. Technical report, 9 2019. DOI: 10.3386/w26280.

[10] Leslie C. Banes, Rachel M. Restani, Rebecca C. Ambrose, Heather A. Martin, and Robert Bayley. Relating performance on written assessments to features of mathematics discussion. *International Journal of Science and Mathematics Education*, 11 2019. [Online; accessed 2020-01-21].

[11] Dan Battey and Luis A. Leyva. A framework for understanding whiteness in mathematics education. *Journal of Urban Mathematics Education*, 9(2):49–80, 2016.

[12] Dan Battey, Luis A. Leyva, Immanuel Williams, Victoria A. Belizario, Rachel Greco, and Roshni Shah. Racial (mis)match in middle school mathematics classrooms: Relational interactions as a racialized mechanism. *Harvard Educational Review*, 88(4):455–482, 12 2018.

[13] Dan Battey, Rebecca A. Neal, Luis Leyva, and Karlyn Adams-Wiggins. The interconnectedness of relational and content dimensions of quality instruction: Supportive teacher–student relationships in urban elementary mathematics classrooms. *The Journal of Mathematical Behavior*, 42:1–19, 6 2016.

[14] D. Baynton. Disability and the justification of inequality in american history. In P. Longmore and L. Umansky, editors, *The new disability history: American perspectives*, pages 33–57. New York University Press, New York, NY, 2001.

[15] Sian L. Beilock. Math performance in stressful situations. *Current Directions in Psychological Science*, 17(5):339–343, 10 2008. publisher: SAGE Publications Inc.

[16] Joseph Berger, Bernard P. Cohen, and Morris Zelditch. Status characteristics and social interaction. *American Sociological Review*, 37(3):241–255, 1972.

[17] R. Q. Berry, Mark Ellis, and Sherick Hughes. Examining a history of failed reforms and recent stories of success: mathematics education and black learners of mathematics in the united states. *Race Ethnicity and Education*, 17(4):540–568, 8 2014. publisher: Routledge.

[18] Kiran Bisra, Qing Liu, John C. Nesbit, Farimah Salimi, and Philip H. Winne. Inducing self-explanation: a meta-analysis. *Educational Psychology Review*, 30(3):703–725, 9 2018.

[19] P. Black and D. Wiliam. Assessment and classroom learning. *Assessment in Education: Principles, Policy & Practice*, 5(1):7–74, 3 1998.

[20] P. Black and D. Wiliam. Developing the theory of formative assessment. *Educational Assessment, Evaluation and Accountability*, 21(1):5–31, 2009.

[21] L. S Blackwell, K. H Trzesniewski, and C. S Dweck. Implicit theories of intelligence predict achievement across an adolescent transition: A longitudinal study and an intervention. *Child Development*, 78(1):246–263, 2007.

[22] S. Blum, editor. *Ungrading: Why Rating Students Undermines Learning (and What to Do Instead)*. West Virginia University Press, Morgantown, 2020.

[23] Güler Boyraz, J. Brandon Waits, and Victoria A. Felix. Authenticity, life satisfaction, and distress: A longitudinal analysis. *Journal of Counseling Psychology*, 61(3):498–505, 2014.

[24] Neil A. Bradbury. Attention span during lectures: 8 seconds, 10 minutes, or more? *Advances in Physiology Education*, 40(4):509–513, 12 2016.

[25] J. Bransford. *How people learn: Brain, mind, experience, and school, Chapter 9: Technology to support learning*. National Academies Press, 2000.

[26] Mary E. Brenner, Richard E. Mayer, Bryan Moseley, Theresa Brar, Richard Durán, Barbara Smith Reed, and David Webb. Learning by understanding: The role of multiple representations in learning algebra. *American Educational Research Journal*, 34(4):663–689, 1 1997. publisher: American Educational Research Association.

[27] David M. Bressoud. The strange role of calculus in the united states. *ZDM*, 2020. [Online; accessed 2020-09-08].

[28] Derek C. Briggs, Maria Araceli Ruiz-Primo, Erin Furtak, Lorrie Shepard, and Yue Yin. Meta-analytic methodology and inferences about the efficacy of formative assessment. *Educational Measurement: Issues and Practice*, 31(4):13–17, 2012.

[29] Susan M. Brookhart. *How to Give Effective Feedback to Your Students*. ASCD, 3 2017. Google-Books-ID: uttODgAAQBAJ.

[30] Amy B. Brunell, Michael H. Kernis, Brian M. Goldman, Whitney Heppner, Patricia Davis, Edward V. Cascio, and Gregory D. Webster. Dispositional authenticity and romantic relationship functioning. *Personality and Individual Differences*, 48(8):900–905, 6 2010.

[31] Anthony S. Bryk, Louis M. Gomez, and Alicia Grunow. Getting ideas into action: Building networked improvement communities in education. In M. T. Hallinan, editor, *Frontiers in sociology of education*, page 127–162. Springer, New York, NY, 2011. [Online; accessed 2014-06-24].

[32] Diane M. Bunce, Elizabeth A. Flens, and Kelly Y. Neiles. How long can students pay attention in class? a study of student attention decline using clickers. *Journal of Chemical Education*, 87(12):1438–1443, 12 2010. publisher: American Chemical Society.

[33] R. Butler. Enhancing and undermining intrinsic motivation: The effects of task-involving and ego-involving evaluation on interest and performance. *British Journal of Educational Psychology*, 58(1):1–14, 1988.

[34] Susanna Calkins, Sharisse Grannan, and Jason Siefken. Using peer-assisted reflection in math to foster critical thinking and communication skills. *PRIMUS*, 6 2019. [Online; accessed 2019-06-12].

[35] Robert Campbell, David Clark, and Jessica OShaughnessy. Introduction to the special issue on implementing mastery grading in the undergraduate mathematics classroom. *PRIMUS*, 30(8-10):837–848, 11 2020.

[36] Joanne Caniglia. Promoting self-reflection over re-teaching: Addressing students' misconceptions with 'my favorite no'. *JME (Journal of Mathematics Education)*, 5(2):70–78, 12 2020. number: 2.

[37] Sylvia Carlisle. Simple specifications grading. *PRIMUS*, 30(8-10):926–951, 11 2020.

[38] A. Cawley, R. Aniceto, S. Ricarte, and R. Wilson. Well, it's obvious": Students' experiences with mathematical microaggressions. In *Proceedings of the 25th Annual Conference for the Research in Undergraduate Mathematics Education*, Omaha, NE, 2023.

[39] S. H. Chapin and C. O'Connor. Academically productive talk: Supporting students' learning in mathematics. *The learning of mathematics*, 69:113–128, 2007.

[40] Sapna Cheryan, Victoria C. Plaut, Paul G. Davies, and Claude M. Steele. Ambient belonging: How stereotypical cues impact gender participation in computer science. *Journal of Personality and Social Psychology*, 97(6):1045–1060, 2009. publisher-place: US publisher: American Psychological Association.

[41] Raj Chetty, John N. Friedman, Emmanuel Saez, Nicholas Turner, and Danny Yagan. Mobility report cards: The role of colleges in intergenerational mobility. Technical report, 2017.

[42] M.T.H. Chi, N. De Leeuw, M.H. Chiu, and C. LaVancher. Eliciting self-explanations improves understanding. *Cognitive science*, 18(3):439–477, 1994.

[43] Deborah Chinn and Claire Homeyard. Easy read and accessible information for people with intellectual disabilities: Is it worth it? a meta-narrative literature review. *Health Expectations*, 20(6):1189–1200, 2017.

[44] Sumi Cho, Kimberlé Williams Crenshaw, and Leslie McCall. Toward a field of intersectionality studies: Theory, applications, and praxis. *Signs: Journal of Women in Culture and Society*, 38(4):785–810, 6 2013. publisher: The University of Chicago Press.

[45] Kerry Chávez and Kristina M. W. Mitchell. Exploring bias in student evaluations: Gender, race, and ethnicity. *PS: Political Science & Politics*, 53(2):270–274, 4 2020. publisher: Cambridge University Press.

[46] Emily Cilli-Turner, Justin Dunmyre, Thomas Mahoney, and Chad Wiley. Mastery grading: Build-a-syllabus workshop. *PRIMUS*, 30(8-10):952–978, 11 2020.

[47] Pauline Rose Clance and Suzanne Ament Imes. The imposter phenomenon in high achieving women: Dynamics and therapeutic intervention. *Psychotherapy: Theory, Research & Practice*, 15(3):241–247, 1978.

[48] P. Cobb, J. Confrey, A. Disessa, R. Lehrer, and L. Schauble. Design experiments in educational research. *Educational researcher*, 32(1):9–13, 2003.

[49] E. G Cohen and R. A Lotan. *Working for equity in heterogeneous classrooms: Sociological theory into practice.* Teachers College Press, New York, NY, 1997.

[50] E. G. Cohen and R. A. Lotan. *Designing Groupwork: Strategies for the Heterogeneous Classroom Third Edition*. Teachers College Press, 6 2014. Google-Books-ID: NW7lAwAAQBAJ.

[51] E. G. Cohen and Susan S. Roper. Modification of interracial interaction disability: An application of status characteristic theory. *American Sociological Review*, 37(6):643–657, 1972.

[52] Elizabeth G. Cohen and Rachel A. Lotan. Producing equal-status interaction in the heterogeneous classroom. *American Educational Research Journal*, 32(1):99–120, 3 1995. publisher: American Educational Research Association.

[53] Elizabeth G. Cohen, Rachel A. Lotan, and Chaub Leechor. Can classrooms learn? *Sociology of Education*, 62(2):75–94, 1989. publisher: [Sage Publications, Inc., American Sociological Association].

[54] Elizabeth G. Cohen, Rachel A. Lotan, Beth A. Scarloss, and Adele R. Arellano. Complex instruction: Equity in cooperative learning classrooms. *Theory Into Practice*, 38(2):80–86, 1999. publisher: Taylor & Francis, Ltd.

[55] Geoffrey L. Cohen, Claude M. Steele, and Lee D. Ross. The Mentor's Dilemma: Providing Critical Feedback Across the Racial Divide. *Personality and Social Psychology Bulletin*, 25(10):1302–1318, October 1999. Publisher: SAGE Publications Inc.

[56] P. H. Collins. Black feminist thought in the matrix of domination. In *Black feminist thought: Knowledge, consciousness, and the politics of empowerment*, pages 221–238. 1990.

[57] Katherine Connort. The legal implications of gender bias in standardized testing. *Berkeley Women's Law Journal*, 7(1):13–89, 1992.

[58] Avraham Z. Cooper and Jeremy B. Richards. Lectures for adult learners: Breaking old habits in graduate medical education. *The American Journal of Medicine*, 130(3):376–381, 3 2017.

[59] Katelyn M. Cooper and Sara E. Brownell. Coming out in class: Challenges and benefits of active learning in a biology classroom for lgbtqia students. *CBE—Life Sciences Education*, 15(3):ar37, 9 2016.

[60] Katelyn M. Cooper, Jeffrey N. Schinske, and Kimberly D. Tanner. Reconsidering the share of a think–pair–share: Emerging limitations, alternatives, and opportunities for research. *CBE—Life Sciences Education*, 20(1):fe1, 1 2021. publisher: American Society for Cell Biology (lse).

[61] Common Core. Common core state standards for mathematics. Technical report, Washington, DC, 2010.

[62] National Research Council. *Adding it up: Helping children learn mathematics*. Mathematics Learning Study Committee, Center for Education, Division of Behavioral and Social Sciences and Education. National Academy Press, Washington, DC, 2001.

[63] Bryan R. Crissinger. The effect of distributed practice in undergraduate statistics homework sets: A randomized trial. *Journal of Statistics Education*, 23(3):null, 11 2015. publisher: Taylor & Francis.

[64] A. Cuoco, E. P. Goldenberg, and J. Mark. Habits of mind: An organizing principle for mathematics curricula. *Journal of Mathematical Behavior*, 15:375–402, 1996.

[65] Diane Dalby and Malcolm Swan. Using digital technology to enhance formative assessment in mathematics classrooms. *British Journal of Educational Technology*, 50(2):832–845, 3 2019. publisher: John Wiley & Sons, Ltd.

[66] Nilanjana Dasgupta, Melissa McManus Scircle, and Matthew Hunsinger. Female peers in small work groups enhance women's motivation, verbal participation, and career aspirations in engineering. *Proceedings of the National Academy of Sciences*, 112(16):4988–4993, 4 2015. PMID: 25848061.

[67] E. L Deci and R. M Ryan. Self-determination theory: A macrotheory of human motivation, development, and health. *Canadian Psychology/Psychologie canadienne*, 49(3):182, 2008.

[68] Peter Demerath. *Producing success: The culture of personal advancement in an American high school.* University of Chicago Press, Chicago, IL, December 2009.

[69] Christina DeRoche. Labels in education: The role of parents and parental cultural capital in acquiring diagnoses and educational accommodations. *Canadian Journal of Education / Revue canadienne de l'éducation*, 38(4):1–24, 2015. Publisher: Canadian Society for the Study of Education.

[70] Louis Deslauriers, Logan S. McCarty, Kelly Miller, Kristina Callaghan, and Greg Kestin. Measuring actual learning versus feeling of learning in response to being actively engaged in the classroom. *Proceedings of the National Academy of Sciences*, page 201821936, 9 2019. PMID: 31484770.

[71] A. A diSessa. Knowledge in pieces. In G. Forman and P. Pufall, editors, *Constructivism in the Computer Age*, pages 49–61. Lawrence Erlbaum Associates, Hillsdale, NJ, 1988.

[72] Anika Dreher and Sebastian Kuntze. Teachers' professional knowledge and noticing: The case of multiple representations in the mathematics classroom. *Educational Studies in Mathematics*, 88(1):89–114, 1 2015.

[73] John Dunlosky, Katherine A. Rawson, Elizabeth J. Marsh, Mitchell J. Nathan, and Daniel T. Willingham. Improving students' learning with effective learning techniques: Promising directions from cognitive and educational psychology. *Psychological Science in the Public Interest*, 14(1):4–58, 1 2013. publisher: SAGE Publications Inc.

[74] D. Dunning, K. Johnson, J. Ehrlinger, and J. Kruger. Why people fail to recognize their own incompetence. *Current Directions in Psychological Science*, 12(3):83, 2003.

[75] Brianne Dávila. Critical race theory, disability microaggressions and latina/o student experiences in special education. *Race Ethnicity and Education*, 18(4):443–468, 7 2015. publisher: Routledge.

[76] Carol Dweck. *Mindset: The new psychology of success.* Random House, New York, NY, 2006. [Online; accessed 2015-02-18].

[77] Carol S. Dweck and Ellen L. Leggett. A social-cognitive approach to motivation and personality. *Psychological review*, 95(2):256, 1988. Publisher: American Psychological Association.

[78] Jessica Brooke Ernest, Daniel L. Reinholz, and Niral Shah. Hidden competence: women's mathematical participation in public and private classroom spaces. *Educational Studies in Mathematics*, 102(2):153–172, 10 2019.

[79] P. Ernest. The nature of mathematics: Towards a social constructivist account. *Science & Education*, 1(1):89–100, 3 1992.

[80] P. Ernest. The popular image of mathematics. *Humanistic Mathematics Network Journal*, 1(8):53–55, 1992.

[81] Indigo Esmonde and Jennifer M. Langer-Osuna. Power in numbers: Student participation in mathematical discussions in heterogeneous spaces. *Journal for Research in Mathematics Education*, 44(1):288–315, 2013.

[82] Nancy Falchikov. *Learning together: Peer tutoring in higher education.* Routledge/Falmer, London, 2003. [Online; accessed 2013-05-14].

[83] Y. Fan, L. J. Shepherd, E. Slavich, D. Waters, M. Stone, R. Abel, and E. L. Johnston. Gender and cultural bias in student evaluations: Why representation matters. *PLOS ONE*, 14(2):e0209749, 2 2019. publisher: Public Library of Science.

[84] James Farley, Evan Risko, and Alan Kingstone. Everyday attention and lecture retention: the effects of time, fidgeting, and mind wandering. *Frontiers in Psychology*, 4, 2013. [Online; accessed 2022-09-19].

[85] Andi Mulawakkan Firdaus. Application of cooperative learning model type think pair share (tps) on mathematical communication ability. *Daya Matematis: Jurnal Inovasi Pendidikan Matematika*, 7(1):59–68, 4 2019. number: 1.

[86] Paulette C. Flore and Jelte M. Wicherts. Does stereotype threat influence performance of girls in stereotyped domains? A meta-analysis. *Journal of School Psychology*, 53(1):25–44, February 2015.

[87] M. Foucault. *Discipline and punish: The birth of the prison.* Pantheon Books, New York, NY, 1977. [Online; accessed 2015-07-08].

[88] R. Freedle. How and why standardized tests systematically underestimate african-americans' true verbal ability and what to do about it: Towards the promotion of two new theories with practical implications. *St. John's Law Review*, 80:183–225, 2006.

[89] Scott Freeman, Sarah L. Eddy, Miles McDonough, Michelle K. Smith, Nnadozie Okoroafor, Hannah Jordt, and Mary Pat Wenderoth. Active learning increases student performance in science, engineering, and mathematics. *Proceedings of the National Academy of Sciences*, 111(23):8410–8415, 5 2014. PMID: 24821756.

[90] P. Freire. *Pedagogy of the Oppressed.* Herder and Herder, New York, NY, 1970.

[91] Sharon Fries-Britt and Jeanette Snider. Mentoring outside the line: The importance of authenticity, transparency, and vulnerability in effective mentoring relationships. *New Directions for Higher Education*, 2015(171):3–11, 2015.

[92] R. E Fullilove and P. U Treisman. Mathematics achievement among african american undergraduates at the university of california, berkeley: An evaluation of the mathematics workshop program. *The Journal of Negro Education*, 59(3):463–478, 1990.

[93] Nichole M. Garcia, Nancy López, and Verónica N. Vélez. Quantcrit: rectifying quantitative methods through critical race theory. *Race Ethnicity and Education*, 21(2):149–157, 3 2018.

[94] Maisie L. Gholson. Clean corners and algebra: A critical examination of the constructed invisibility of black girls and women in mathematics. *The Journal of Negro Education*, 85(3):290–301, 2016. publisher: Journal of Negro Education.

[95] Henry A. Giroux and Anthony N. Penna. Social education in the classroom: The dynamics of the hidden curriculum. *Theory & Research in Social Education*, 7(1):21–42, 3 1979.

[96] Tolga Gok. The evaluation of conceptual learning and epistemological beliefs on physics learning by think-pair-share. *Journal of Education in Science Environment and Health*, 4(1):69–80, 1 2018. number: 1.

[97] Rosemary Gonzalez and Christina U. Ayala-Alcantar. Critical caring: Dispelling latino stereotypes among preservice teachers. *Journal of Latinos and Education*, 7(2):129–143, 3 2008.

[98] Catherine Good, Aneeta Rattan, and Carol S. Dweck. Why do women opt out? sense of belonging and women's representation in mathematics. *Journal of Personality and Social Psychology*, 102(4):700–717, 2012.

[99] Merrilyn Goos, Peter Galbraith, and Peter Renshaw. Socially mediated metacognition: Creating collaborative zones of proximal development in small group problem solving. *Educational Studies in Mathematics*, 49(2):193–223, 2 2002.

[100] Maithreyi Gopalan and Shannon T. Brady. College students' sense of belonging: A national perspective. *Educational Researcher*, 49(2):134–137, 3 2020. publisher: American Educational Research Association.

[101] B. Grossen and S. Ewing. Raising mathematics problem-solving performance: Do the NCTM teaching standards help? Final report. *Effective School Practices*, 13(2):79–91, 1994.

[102] Rochelle Gutiérrez. A "gap-gazing" fetish in mathematics education? problematizing research on the achievement gap. *Journal for Research in Mathematics Education*, 39(4):357–364, 7 2008.

[103] Gila Hanna and E. Barbeau. Proofs as bearers of mathematical knowledge. *ZDM*, 40(3):345–353, 2008.

[104] G. Harel and L. Sowder. Toward comprehensive perspectives on the learning and teaching of proof. In F. Lester, editor, *Second handbook of research on mathematics teaching and learning*, page 805–842. National Council of Teachers of Mathematics, Charlotte, NC, 2007.

[105] Pamela E. Harris, Alicia Prieto Langarica, Vanessa Rivera Quiñones, Luis Alfonso Sordo Vieira, Rosaura Uscanga Lomeli, Andrés Rodolfo Vindas Meléndez, and Ana Valle, editors. *Testimonios: stories of Latinx and Hispanic mathematicians.* Classroom resource materials. American Mathematical Society/Mathematical Association of America, Providence, Rhode Island, 2021.

[106] F. Harris III and L. Wood. How to respond to racial microaggressions when they occur, 5 2020. [Online; accessed 2022-02-15].

[107] J. Hattie and H. Timperley. The power of feedback. *Review of educational research*, 77(1):81–112, 2007.

[108] Leslie R. M. Hausmann, Janet Ward Schofield, and Rochelle L. Woods. Sense of belonging as a predictor of intentions to persist among african american and white first-year college students. *Research in Higher Education*, 48(7):803–839, 11 2007.

[109] Kaleen Healey and Chloe Stroman. Structures for belonging: A synthesis of research on belonging-supportive learning environments. Technical report, 2021.

[110] Michael Henderson, Michael Phillips, Tracii Ryan, David Boud, Phillip Dawson, Elizabeth Molloy, and Paige Mahoney. Conditions that enable effective feedback. *Higher Education Research & Development*, 38(7):1401–1416, 11 2019.

[111] A. K. Henrich, E. D. Lawrence, M. A. Pons, and D. G. Taylor, editors. *Living proof: Stories of resilience along the mathematical journey.* MAA and AMS, 2019.

[112] Joan Herman, Scott Epstein, Seth Leon, Deborah La Torre Matrundola, Sarah Reber, and Kilchan Choi. Implementation and effects of ldc and mdc in kentucky districts. Technical report, Los Angeles, 2014. [Online; accessed 2015-02-04].

[113] Reuben Hersh. *What is mathematics, really?* Oxford University Press, Oxford, UK, 1997.

[114] Hetika Hetika, Ida Farida, and Yeni Priatna Sari. Think pair share (tps) as method to improve student's learning motivation and learning achievement. *Dinamika Pendidikan*, 12(2):125–135, 12 2017. number: 2.

[115] H.C. Hill, B. Rowan, and D.L. Ball. Effects of teachers' mathematical knowledge for teaching on student achievement. *American Educational Research Journal*, 42(2):371, 2005.

[116] W. W. Hoffer. *Minds on mathematics: Using math workshop to develop deep understanding in grades 4-8.* Heinemann, Portsmouth, NH, 2012.

[117] D. Holland, W. Lachicotte, D. Skinner, and C. Cain. *Identity and agency in cultural worlds.* Harvard University Press, Cambridge, MA, 1998.

[118] bell hooks. *Teaching to Transgress: Education as the Practice of Freedom.* Routledge, New York, 9 1994.

[119] E. Hsu, T. J. Murphy, U. Treisman, M. P. Carlson, and C. Rasmussen. Supporting high achievement in introductory mathematics courses: What we have learned from 30 years of the emerging scholars program. In *Making the connection: Research and teaching in undergraduate mathematics education*, volume 73 of *Maa Notes*, page 205. 2008. [Online; accessed 2017-03-03].

[120] Sylvia Hurtado and Deborah Faye Carter. Effects of college transition and perceptions of the campus racial climate on latino college students' sense of belonging. *Sociology of Education*, 70(4):324–345, 10 1997.

[121] Danielle L. Iamarino. The benefits of standards-based grading: A critical evaluation of modern grading practices. *Current Issues in Education*, 17(2), 5 2014. publisher: Arizona State University, Mary Lou Fulton Institute and Graduate School of Education.

[122] Marsha Ing, Noreen M. Webb, Megan L. Franke, Angela C. Turrou, Jacqueline Wong, Nami Shin, and Cecilia H. Fernandez. Student participation in elementary mathematics classrooms: the missing link between teacher practices and student achievement? *Educational Studies in Mathematics*, 90(3):341–356, 2015. publisher: Springer.

[123] Matthew Inglis and Colin Foster. Five decades of mathematics education research. *Journal for Research in Mathematics Education*, 49(4):462–500, 2018.

[124] Sins Invalid. *Skin, tooth, and bone: The basis of movement is our people (2nd ed.)*. Berkeley, CA, 2019.

[125] Amanda Jansen. *Rough Draft Math: Revising to Learn*. Stenhouse Publishers, Portsmouth, New Hampshire, 3 2020.

[126] Lisa M. Jilk. Supporting teacher noticing of students' mathematical strengths. *Mathematics Teacher Educator*, 4(2):188–199, 2016.

[127] Sue Joseph, Carolyn Rickett, Maria Northcote, and Beverly J. Christian. 'who are you to judge my writing?': Student collaboration in the co-construction of assessment rubrics. *New Writing*, 17(1):31–49, 1 2020.

[128] Mahmoud Kaddoura. Think pair share: A teaching learning strategy to enhance students' critical thinking. *Educational Research Quarterly*, 36(4):3–24, 6 2013. publisher: Behavioral Research Press.

[129] Julia Rose Karpicz. "just my being here is self-advocacy": Exploring the self-advocacy experiences of disabled graduate students of color. *JCSCORE*, 6(1):137–163, 7 2020. number: 1.

[130] Leanne R. Ketterlin-Geller. Knowing what all students know: Procedures for developing universal design for assessment. *The Journal of Technology, Learning and Assessment*, 4(2), 11 2005. number: 2.

[131] Peter Kilgour, Maria Northcote, Anthony Williams, and Andrew Kilgour. A plan for the co-construction and collaborative use of rubrics for student learning. *Assessment & Evaluation in Higher Education*, 45(1):140–153, 1 2020.

[132] Alison King, Anne Staffieri, and Anne Adelgais. Mutual peer tutoring: Effects of structuring tutorial interaction to scaffold peer learning. *Journal of Educational Psychology*, 90(1):134–152, 3 1998.

[133] Neal Kingston and Brooke Nash. Formative assessment: A meta-analysis and a call for research. *Educational Measurement: Issues and Practice*, 30(4):28–37, 2011.

[134] Avraham N. Kluger and Angelo DeNisi. The effects of feedback interventions on performance: A historical review, a meta-analysis, and a preliminary feedback intervention theory. *Psychological Bulletin*, 119:254–284, 1996. publisher-place: US publisher: American Psychological Association.

[135] Alison C. Koenka, Lisa Linnenbrink-Garcia, Hannah Moshontz, Kayla M. Atkinson, Carmen E. Sanchez, and Harris Cooper. A meta-analysis on the impact of grades and comments on academic motivation and achievement: a case for written feedback. *Educational Psychology*, 41(7):922–947, 8 2021.

[136] Marina Kogan and Sandra L. Laursen. Assessing long-term effects of inquiry-based learning: A case study from college mathematics. *Innovative Higher Education*, 39(3):183–199, 6 2014.

[137] Rachel Lambert and Paulo Tan. Conceptualizations of students with and without disabilities as mathematical problem solvers in educational research: A critical review. *Education Sciences*, 7(2):51, 6 2017. number: 2 publisher: Multidisciplinary Digital Publishing Institute.

[138] Jennifer M. Langer-Osuna. How brianna became bossy and kofi came out smart: Understanding the trajectories of identity and engagement for two group leaders in a project-based mathematics classroom. *Canadian Journal of Science, Mathematics and Technology Education*, 11(3):207–225, 7 2011.

[139] Jennifer M. Langer-Osuna. The social construction of authority among peers and its implications for collaborative mathematics problem solving. *Mathematical Thinking and Learning*, 18(2):107–124, 4 2016.

[140] Gregory V. Larnell. More than just skill: Examining mathematics identities, racialized narratives, and remediation among black undergraduates. *Journal for Research in Mathematics Education*, 47(3):233–269, 5 2016. publisher: National Council of Teachers of Mathematics section: Journal for Research in Mathematics Education.

[141] Gregory V. Larnell, Denise Boston, and John Bragelman. The stuff of stereotypes: Toward unpacking identity threats amid african american students' learning experiences. *Journal of Education*, 194(1):49–57, 1 2014. publisher: SAGE Publications Inc.

[142] Sandra Laursen, Marja-Liisa Hassi, Marina Kogan, Anne-Barrie Hunter, and Tim Weston. Evaluation of the ibl mathematics project: student and instructor outcomes of inquiry-based learning in college mathematics. *Colorado University*, 2011. [Online; accessed 2014-03-20].

[143] Sandra L. Laursen, Marja-Liisa Hassi, Marina Kogan, and Timothy J. Weston. Benefits for women and men of inquiry-based learning in college mathematics: A multi-institution study. *Journal for Research in Mathematics Education*, 45(4):406–418, 2014.

[144] J. Lave and E. Wenger. *Situated learning: Legitimate peripheral participation*. Cambridge University Press, Cambridge, MA, 1991.

[145] Hansol Lee, Huy Q. Chung, Yu Zhang, Jamal Abedi, and Mark Warschauer. The effectiveness and features of formative assessment in us k-12 education: A systematic review. *Applied Measurement in Education*, 33(2):124–140, 4 2020.

[146] Meggan J. Lee, Jasmine D. Collins, Stacy Anne Harwood, Ruby Mendenhall, and Margaret Browne Huntt. "if you aren't white, asian or indian, you aren't an engineer": racial microaggressions in stem education. *International Journal of STEM Education*, 7(1):48, 9 2020.

[147] Colleen M. Lewis and Niral Shah. How equity and inequity can emerge in pair programming. ICER '15, page 41–50, New York, NY, USA, 8 2015. Association for Computing Machinery. [Online; accessed 2022-09-15].

[148] K. L. Lewis, J. G. Stout, S. J. Pollock, N. Finkelstein, and T. A. Ito. Fitting in or opting out: A review of key social-psychological factors influencing a sense of belonging for women in physics. *Physical Review Physics Education Research*, 12(2):020110, 8 2016.

[149] Luis Leyva, Taylor McNeill, Brittany Marshall, and Oscar Guzmán. "it seems like they purposefully try to make as many kids drop": An analysis of logics and mechanisms of racial-gendered inequality in introductory mathematics instruction. *The Journal of Higher Education*, 3 2021.

[150] Hongli Li, Yao Xiong, Charles Vincent Hunter, Xiuyan Guo, and Rurik Tywoniw. Does peer assessment promote student learning? a meta-analysis. *Assessment & Evaluation in Higher Education*, 45(2):193–211, 2 2020.

[151] Lingyu Li, Helena Lourdes Donato-Sapp, Nirmala Erevelles, Lisette E. Torres, and Federico Waitoller. A kitchen-table talk against ableism: Disability justice for collective liberation. *Equity & Excellence in Education*, 0(0):1–14, 3 2022.

[152] Wei Liu. Does teacher immediacy affect students? a systematic review of the association between teacher verbal and non-verbal immediacy and student motivation. *Frontiers in Psychology*, 12, 2021. [Online; accessed 2022-09-14].

[153] Paul Lockhart. *A Mathematician's Lament: How School Cheats Us Out of Our Most Fascinating and Imaginative Art Form.* Bellevue Literary Press, 4 2009.

[154] Beatrice Lok, Carmel McNaught, and Kenneth Young. Criterion-referenced and norm-referenced assessments: compatibility and complementarity. *Assessment & Evaluation in Higher Education*, 41(3):450–465, 4 2016.

[155] R. Lotan. Group-worthy tasks. *Educational Leadership*, 60(6):72–75, 2003.

[156] R. A. Lotan. Equitable classrooms: A compelling connection between theory and practice. In M. Webster and L. S. Walker, editors, *Unequals*, pages 178–199. Oxford University Press, 5 2022. DOI: 10.1093/oso/9780197600009.003.0009.

[157] Nicole L. Louie. The culture of exclusion in mathematics education and its persistence in equity-oriented teaching. *Journal for Research in Mathematics Education*, 48(5):488–519, 11 2017. publisher: National Council of Teachers of Mathematics section: Journal for Research in Mathematics Education.

[158] B. Love. *We want to do more than survive: Abolitionist teaching and the pursuit of educational freedom.* Beacon Press, Boston, MA, 2019.

[159] S. T. Lubienski and C. M. Ganley. Research on gender and mathematics. In J. Cai, editor, *Compendium for Research in Mathematics Education*, pages 649–666. National Council of Teachers of Mathematics, Reston, VA, 2017.

[160] J. Malkevitch. Mathematics research projects, 2022. [Online; accessed 2022-06-02].

[161] D. B. Martin. Learning mathematics while black. *Educational Foundations*, 26(1-2):47–66, 2012. publisher: Caddo Gap Press.

[162] D. B. Martin. Equity, inclusion, and antiblackness in mathematics education. *Race Ethnicity and Education*, 22(4):459–478, 7 2019.

[163] D. B. Martin, C. Rousseau-Anderson, and N. Shah. Race and mathematics education. In J. Cai, editor, *Compendium for Research in Mathematics Education*, pages 607–636. National Council of Teachers of Mathematics, Reston, VA, 2017.

[164] David M. Marx and Jasmin S. Roman. Female role models: Protecting women's math test performance. *Personality and Social Psychology Bulletin*, 28(9):1183–1193, 9 2002. publisher: SAGE Publications Inc.

[165] Allison Master, Sapna Cheryan, and Andrew N. Meltzoff. Computing whether she belongs: Stereotypes undermine girls' interest and sense of belonging in computer science. *Journal of Educational Psychology*, 108(3):424–437, 2016.

[166] Myosha McAfee. The kinesiology of race. *Harvard Educational Review*, 84(4):468–491, 12 2014.

[167] Jan McArthur. Assessment for social justice: the role of assessment in achieving social justice. *Assessment & Evaluation in Higher Education*, 41(7):967–981, 10 2016.

[168] Ebony Omotola McGee. Interrogating structural racism in stem higher education. *Educational Researcher*, 49(9):633–644, 12 2020. publisher: American Educational Research Association.

[169] Hazel J. McKenna, Kuo-Liang Chang, and Thomas Mgonja. A framework to measure microaggressions in the mathematics classroom. *SN Social Sciences*, 1(5):122, 4 2021.

[170] James H. McMillan, Jessica C. Venable, and Divya Varier. Studies of the effect of formative assessment on student achievement: So much more is needed. *Practical Assessment, Research & Evaluation*, 18(2), 2 2013. publisher: Center for Educational Assessment.

[171] Katherine L. McNeill and Joseph Krajcik. Scientific explanations: Characterizing and evaluating the effects of teachers' instructional practices on student learning. *Journal of Research in Science Teaching*, 45(1):53–78, 2008.

[172] Hugh Mehan. "what time is it, denise?": Asking known information questions in classroom discourse. *Theory into Practice*, 18(4):285–294, 10 1979.

[173] Kathleen Melhuish, Eva Thanheiser, Alexander White, Brenda Rosencrans, J. Michael Shaughnessy, Linda Foreman, Andrew Riffel, and Layla Guyot. The efficacy of research-based "mathematics for all" professional development. *Journal for Research in Mathematics Education*, 53(4):307–333, 7 2022. publisher: National Council of Teachers of Mathematics section: Journal for Research in Mathematics Education.

[174] Mia Mingus. Moving toward the ugly: A politic beyond desirability, 8 2011. [Online; accessed 2022-01-13].

[175] Alison Mirin and Paulo Tan. Disability accommodations in college: Alarming discrimination in mathematics. volume Proceedings of The 24th Annual Conference on Research in Undergraduate Mathematics Education (RUME), Boston, MA, 2 2022.

[176] Katja Mäntylä. Idioms and language users : the effect of the characteristics of idioms on their recognition and interpretation by native and non-native speakers of english. *Jyväskylä studies in humanities*, (13), 2004.

[177] Elizabeth Molloy, David Boud, and Michael Henderson. Developing a learning-centred framework for feedback literacy. *Assessment & Evaluation in Higher Education*, 45(4):527–540, 5 2020.

[178] Jennifer Morrow and Margot Ackermann. Intention to persist and retention of first-year students: The importance of motivation and sense of belonging. *College Student Journal*, 46(3):483–491, 9 2012.

[179] Jason Morton, Maria Northcote, Peter Kilgour, and Wendy Jackson. Sharing the construction of assessment rubrics with students: A model for collaborative rubric construction. *Journal of University Teaching & Learning Practice*, 18(4), 7 2021.

[180] Corinne A. Moss-Racusin, John F. Dovidio, Victoria L. Brescoll, Mark J. Graham, and Jo Handelsman. Science faculty's subtle gender biases favor male students. *Proceedings of the National Academy of Sciences*, 109(41):16474–16479, 10 2012. PMID: 22988126.

[181] Ann L. Mullen. Elite destinations: pathways to attending an ivy league university. *British Journal of Sociology of Education*, 30(1):15–27, 1 2009.

[182] Joel T. Nadler and M. H. Clark. Stereotype Threat: A Meta-Analysis Comparing African Americans to Hispanic Americans. *Journal of Applied Social Psychology*, 41(4):872–890, 2011. _eprint: https://onlinelibrary.wiley.com/doi/pdf/10.1111/j.1559-1816.2011.00739.x.

[183] Francis Sahngun Nahm. What the P values really tell us. *The Korean Journal of Pain*, 30(4):241–242, October 2017.

[184] N. S. Nasir. Identity, goals, and learning: Mathematics in cultural practice. *Mathematical Thinking and Learning*, 4(2-3):213–247, 2002.

[185] N. S. Nasir, V. Hand, and E. V. Taylor. Culture and mathematics in school: Boundaries between "cultural" and "domain" knowledge in the mathematics classroom and beyond. *Review of Research in Education*, 32(1):187–240, 2008.

[186] Katharina Barzagar Nazari and Mirjam Ebersbach. Distributed practice in mathematics: Recommendable especially for students on a medium performance level? *Trends in Neuroscience and Education*, 17:100122, 12 2019.

[187] NCTM. *Curriculum and evaluation standards for school mathematics*. National Council of Teachers of Mathematics: Commission on Standards for School Mathematics, 1989.

[188] Hannah-Hanh D. Nguyen and Ann Marie Ryan. Does stereotype threat affect test performance of minorities and women? a meta-analysis of experimental evidence. *Journal of Applied Psychology*, 93(6):1314–1334, 2008.

[189] Yujing Ni, De-Hui Ruth Zhou, Jinfa Cai, Xiaoqing Li, Qiong Li, and Iris X. Sun. Improving cognitive and affective learning outcomes of students through mathematics instructional tasks of high cognitive demand. *The Journal of Educational Research*, 111(6):704–719, 11 2018.

[190] Juuso Henrik Nieminen. Student conceptions of assessment accommodations in university mathematics: An analysis of power. *Nordic Studies in Mathematics Education*, 25(3-4):27–49, 2020.

[191] Juuso Henrik Nieminen. Governing the 'disabled assessee': a critical reframing of assessment accommodations as sociocultural practices. *Disability & Society*, 0(0):1–28, 2 2021.

[192] Juuso Henrik Nieminen. Assessment for inclusion: rethinking inclusive assessment in higher education. *Teaching in Higher Education*, 0(0):1–19, 1 2022.

[193] Juuso Henrik Nieminen. Unveiling ableism and disablism in assessment: a critical analysis of disabled students' experiences of assessment and assessment accommodations. *Higher Education*, 5 2022. [Online; accessed 2022-06-13].

[194] Juuso Henrik Nieminen, Man Ching Esther Chan, and David Clarke. What affordances do open-ended real-life tasks offer for sharing student agency in collaborative problem-solving? *Educational Studies in Mathematics*, 109(1):115–136, 1 2022.

[195] Juuso Henrik Nieminen and Henri Valtteri Pesonen. Politicising inclusive learning environments: how to foster belonging and challenge ableism? *Higher Education Research & Development*, 0(0):1–14, 7 2021.

[196] Linda B. Nilson and Claudia J. Stanny. *Specifications Grading: Restoring Rigor, Motivating Students, and Saving Faculty Time.* Stylus Publishing, Sterling, Virginia, reprint edition edition, 10 2014.

[197] Y. Ningsih, R. Andika, I. K. Sari, S. Ahmad, and A. K. Kenedi. The application think pair share in learning mathematic. *Journal of Physics: Conference Series*, 1321(2):022092, 10 2019. publisher: IOP Publishing.

[198] Rafael E. Núñez. No innate number line in the human brain. *Journal of Cross-Cultural Psychology*, 42(4):651–668, 5 2011. publisher: SAGE Publications Inc.

[199] Jeannie Oakes. Multiplying inequalities: The effects of race, social class, and tracking on opportunities to learn mathematics and science. Technical report, Santa Monica, CA, 1990. [Online; accessed 2013-11-22].

[200] C. O'Connor, S. Michaels, and S. Chapin. "scaling down" to explore the role of talk in learning: From district intervention to controlled classroom study. In *Socializing intelligence through academic talk and dialogue*, pages 111–126. American Educational Research Association, 2015.

[201] President's Council of Advisors on Science and Technology. Engage to excel: Producing one million additional college graduates with degrees in science, technology, engineering, and mathematics. Technical report, Washington, D.C., 2012. [Online; accessed 2013-10-29].

[202] Mathematical Association of America. Instructional practices guide. Technical report, Washington, D.C., 2018.

[203] National Council of Teachers of Mathematics. *Principles and standards for school mathematics.* The National Council of Teachers of Mathematics, Reston, VA, 2000.

[204] National Council of Teachers of Mathematics. Principles to actions: Ensuring mathematics success for all. Technical report, Reston, VA, 2014.

[205] Conference Board of the Mathematical Sciences. Active learning in post-secondary mathematics education. Technical report, Washington, DC, 2016.

[206] Cathy O'Neil. *Weapons of math destruction: How big data increases inequality and threatens democracy.* Crown, New York, NY, 2016.

[207] Joan M Ostrove and Susan M Long. Social class and belonging: Implications for college adjustment. *The Review of Higher Education*, 30(4):363–389, 2007. publisher: Johns Hopkins University Press.

[208] K. Owens. A beginner's guide to standards based grading |, 2015. [Online; accessed 2022-06-13].

[209] Anna Parkman. The imposter phenomenon in higher education: Incidence and impact. *Journal of Higher Education Theory and Practice*, 16(1):51, 2016.

[210] Frederick Peck. Towards anti-deficit education in undergraduate mathematics education: How deficit perspectives work to structure inequality and what can be done about it. *PRIMUS*, 0(0):1–22, 6 2020.

[211] Mariel A. Pfeifer, Eve Melanie Reiter, McKenna Hendrickson, and Julie Dangremond Stanton. Speaking up: a model of self-advocacy for stem undergraduates with adhd and/or specific learning disabilities. *International Journal of STEM Education*, 7(1):33, 7 2020.

[212] Jean Piaget. *The principles of genetic epistemology*, volume 7. Routledge & Kegan Paul, London, 1972.

[213] Susan H. Picker and John S. Berry. Investigating pupils' images of mathematicians. *Educational Studies in Mathematics*, 43(1):65–94, 7 2000.

[214] Gavin R. Price, Michèle M. M. Mazzocco, and Daniel Ansari. Why mental arithmetic counts: Brain activation during single digit arithmetic predicts high school math scores. *Journal of Neuroscience*, 33(1):156–163, 1 2013. publisher: Society for Neuroscience section: Articles PMID: 23283330.

[215] Priscillia Pui, Brenda Yuen, and Happy Goh. Using a criterion-referenced rubric to enhance student learning: a case study in a critical thinking and writing module. *Higher Education Research & Development*, 40(5):1056–1069, 7 2021.

[216] Katherine N. Quinn, Michelle M. Kelley, Kathryn L. McGill, Emily M. Smith, Zachary Whipps, and N. G. Holmes. Group roles in unstructured labs show inequitable gender divide. *Physical Review Physics Education Research*, 16(1):010129, 5 2020. publisher: American Physical Society.

[217] J. Rámö. Oppimistavoitematriisi - lineaarialgebra ja matriisilaskenta i, 2022. [Online; accessed 2022-06-13].

[218] D. Kim Reid and Michelle G. Knight. Disability justifies exclusion of minority students: A critical history grounded in disability studies. *Educational Researcher*, 35(6):18–23, 8 2006. publisher: American Educational Research Association.

[219] D. L. Reinholz. Peer-assisted reflection: A design-based intervention for improving success in calculus. *International Journal of Research in Undergraduate Mathematics Education*, 1(2):234–267, 2015.

[220] D. L. Reinholz. The assessment cycle: a model for learning through peer assessment. *Assessment & Evaluation in Higher Education*, 41(2):301–315, 2 2016.

[221] D. L. Reinholz. Improving calculus explanations through peer review. *The Journal of Mathematical Behavior*, 44:34–49, 12 2016.

[222] D. L. Reinholz. Design trees: providing roots for revision in design-based research. *International Journal of Learning Technology*, 12(4):275–293, 1 2017.

[223] D. L. Reinholz. Five practices for supporting inquiry in analysis. *PRIMUS*, 30(1):19–35, 8 2018.

[224] D. L. Reinholz. Reflective apprenticeship for teaching and learning mathematical proof. *Journal of Research in STEM Education*, 4(1):68–80, 2018.

[225] D. L. Reinholz. Disability, mathematics, and the goldilocks conundrum: implications for mathematics education. *For the Learning of Mathematics*, 41(2):19–20, 2021.

[226] D. L. Reinholz, K. Bradfield, and N. Apkarian. Using analytics to support instructor reflection on student participation in a discourse-focused undergraduate mathematics classroom. *International Journal of Research in Undergraduate Mathematics Education*, 5(1):56–74, 4 2019.

[227] D. L. Reinholz, C. Curtis, and N. Shah. Moving beyond gatekeeping: Using data analytics to overcome resistance to pedagogical change. Bolzano, Italy, 2022.

[228] D. L. Reinholz, Estrella Johnson, Christine Andrews-Larson, Amelia Stone-Johnstone, Jessica Smith, Brooke Mullins, Nicholas Fortune, Karen Keene, and N. Shah. When active learning is inequitable: Women's participation predicts gender inequities in mathematical performance. *Journal for Research in Mathematics Education*, 53(3):204–226, 5 2022. publisher: National Council of Teachers of Mathematics section: Journal for Research in Mathematics Education.

[229] D. L. Reinholz, E. Schmied, S. Ridgway, and N. Shah. Utilizing data analytics to incorporate racial equity into stem faculty development. In *Handbook of STEM faculty development*. 2022.

[230] D. L. Reinholz and N. Shah. Equity analytics: A methodological approach for quantifying participation patterns in mathematics classroom discourse. *Journal for Research in Mathematics Education*, 49(2):140–177, 2018.

[231] D. L. Reinholz, A. Stone-Johnstone, and N. Shah. Walking the walk: using classroom analytics to support instructors to address implicit bias in teaching. *International Journal for Academic Development*, 25(3):259–272, 2020.

[232] D. L. Reinholz, A. Stone-Johnstone, I. White, L. M. Sianez, and N. Shah. A pandemic crash course: Learning to teach equitably in synchronous online classes. *CBE—Life Sciences Education*, 19(4):ar60, 12 2020. publisher: American Society for Cell Biology (lse).

[233] Daniel L. Reinholz. Equity during peer conferences: A linguistic comparison by race and gender. *Journal of Research in STEM Education*, 4(1):54–67, 7 2018. number: 1.

[234] Daniel L. Reinholz. Large lecture halls: Whiteboards, not bored students. *PRIMUS*, 0(ja):null, 6 2018.

[235] Daniel L. Reinholz and Mary E. Pilgrim. Student sensemaking of proofs at various distances: the role of epistemic, rhetorical, and ontological distance in the peer review process. *Educational Studies in Mathematics*, 106(2):211–229, 2 2021.

[236] Daniel L. Reinholz, Aileen Reid, and Niral Shah. Not another bias workshop: Using equity analytics to promote antiracist teaching. *Change: The Magazine of Higher Learning*, 54(4):11–17, 7 2022.

[237] Daniel L. Reinholz and Samantha W. Ridgway. Access needs: Centering students and disrupting ableist norms in stem. *CBE—Life Sciences Education*, 20(3):es8, 9 2021. publisher: American Society for Cell Biology (lse).

[238] Daniel L. Reinholz and Anne G. Wilhelm. Race-gender d/discourses in mathematics education: (re)-producing inequitable participation patterns across a diverse, instructionally-advanced urban district. *Urban Education*, pages 1–31, 6 2022. publisher: SAGE Publications Inc.

[239] Marie Agnes Revak. *The effect of distributed practice homework on precalculus achievement at a military academy*. PhD thesis, United States – Florida, 1996. ISBN: 9798209370390.

[240] Evan F. Risko, Nicola Anderson, Amara Sarwal, Megan Engelhardt, and Alan Kingstone. Everyday attention: Variation in mind wandering and memory in a lecture. *Applied Cognitive Psychology*, 26(2):234–242, 2012.

[241] Bethany Rittle-Johnson, Abbey M. Loehr, and Kelley Durkin. Promoting self-explanation to improve mathematics learning: A meta-analysis and instructional design principles. *ZDM*, 49(4):599–611, 8 2017.

[242] Joseph P. Robinson-Cimpian, Sarah Theule Lubienski, Colleen M. Ganley, and Yasemin Copur-Gencturk. Teachers' perceptions of students' mathematics proficiency may exacerbate early gender gaps in achievement. *Developmental Psychology*, 50(4):1262–1281, 2014.

[243] Doug Rohrer, Robert F. Dedrick, and Kaleena Burgess. The benefit of interleaved mathematics practice is not limited to superficially similar kinds of problems. *Psychonomic Bulletin & Review*, 21(5):1323–1330, 10 2014.

[244] Doug Rohrer, Robert F. Dedrick, and Sandra Stershic. Interleaved practice improves mathematics learning. *Journal of Educational Psychology*, 107(3):900–908, 2015. publisher-place: US publisher: American Psychological Association.

[245] Doug Rohrer and Kelli Taylor. The effects of overlearning and distributed practise on the retention of mathematics knowledge. *Applied Cognitive Psychology*, 20(9):1209–1224, 2006.

[246] Rod D. Roscoe and Michelene T. H. Chi. Tutor learning: the role of explaining and responding to questions. *Instructional Science*, 36(4):321–350, 7 2008.

[247] David Rose. Universal design for learning. *Journal of Special Education Technology*, 15(3):45–49, 6 2000. publisher: SAGE Publications Inc.

[248] B. A. Rosenfeld. *A history of non-Euclidean geometry*. Springer, New York, NY, 1988.

[249] M.B. Rowe. Wait time: slowing down may be a way of speeding up! *Journal of teacher education*, 37(1):43–50, 1986.

[250] Maria Araceli Ruiz-Primo, Min Li, Shin-Ping Tsai, and Julie Schneider. Testing one premise of scientific inquiry in science classrooms: Examining students' scientific explanations and student learning. *Journal of Research in Science Teaching*, 47(5):583–608, 2010.

[251] D. Sadker, M. Sadker, and K. R. Zittleman. *Still failing at fairness: How gender bias cheats girls and boys in school and what we can do about it*. Simon and Schuster, New York, NY, 2009. [Online; accessed 2017-03-28].

[252] D.R. Sadler. Formative assessment and the design of instructional systems. *Instructional science*, 18(2):119–144, 1989.

[253] Ariana Sampsel. Finding the effects of think-pair-share on student confidence and participation. *Honors Projects*, 4 2013.

[254] Carrie Sandahl. Queering the crip or cripping the queer?: Intersections of queer and crip identities in solo autobiographical performance. *GLQ: A Journal of Lesbian and Gay Studies*, 9(1):25–56, 2003. publisher: Duke University Press.

[255] Karen Saxe, Linda Braddy, J. Bailer, R. Farinelli, T. Holm, V. Mesa, U. Treisman, and P. Turner. A common vision for undergraduate mathematical science programs in 2025. Technical report, 2015.

[256] A. H. Schoenfeld. *Mathematical problem solving*. Academy Press, New York, 1985. Problem solving consists of one's knowledge base, strategies, metacognition/control, beliefs. To understand problem solving performance, one must consider all of these complementary aspects.

[257] A. H. Schoenfeld. When good teaching leads to bad results: The disasters of 'well-taught' mathematics courses. *Educational psychologist*, 23(2):145–166, 1988.

[258] A. H. Schoenfeld. The math wars. *Educational Policy*, 18(1):253–286, January 2004.

[259] W. G. Secada. Educational equity versus equality of education: An alternative conception. In W. G. Secada, editor, *Equity in education*, page 68–88. Falmer, New York, NY, 1989.

[260] Denise Sekaquaptewa. Gender-based microaggressions in stem settings. *NCID Currents*, 1(1), 2019.

[261] E. Setren, K. Greenberg, O. Moore, and M. Yankovich. Effects of the flipped classroom: Evidence from a randomized trial. Technical report, Cambridge, MA, 2019.

[262] Elaine Seymour and Nancy M. Hewitt. *Talking about leaving: Why undergraduates leave the sciences*, volume 12. Westview Press, Boulder, CO, 1997. [Online; accessed 2014-04-09].

[263] A. Sfard. On two metaphors for learning and the dangers of choosing just one. *Educational Researcher*, 27(2):4–13, 3 1998.

[264] N. Shah. Race, ideology, and academic ability: A relational analysis of racial narratives in mathematics. *Teachers College Record*, 119(7):1–42, 2017.

[265] N. Shah. "asians are good at math" is not a compliment: Stem success as a threat to personhood. *Harvard Educational Review*, 89(4):661–686, 12 2019.

[266] N. Shah, J. A. Christensen, N. A. Ortiz, A. Nguyen, S. Byun, D. Stroupe, and D. L. Reinholz. Racial hierarchy and masculine space: Participatory in/equity in computational physics classrooms. *Computer Science Education*, 0(0):1–25, 9 2020.

[267] N. Shah and S. Crespo. Cultural narratives and status hierarchies. In R. Hunter, M. Civil, B. Herbel-Eisenmann, N. Planas, and D. Wagner, editors, *Mathematical discourse that breaks barriers and creates space for marginalized learners*, pages 23–38. Sense Publishers, Rotterdam, The Netherlands, 2017.

[268] N. Shah, B. Herbel-Eisenmann, and D. L. Reinholz. Why mrs. stone never calls on debra: A case of race-gender ideology in practice. pages 1974–1981, Nashville, TN, 2020. International Society of the Learning Sciences.

[269] Niral Shah, Colleen Lewis, and Roxane Caires. Analyzing equity in collaborative learning situations: A comparative case study in elementary computer science. 6 2014. publisher: Boulder, CO: International Society of the Learning Sciences.

[270] V.J. Shute. Focus on formative feedback. *Review of Educational Research*, 78(1):153–189, 2008.

[271] Robert E. Slavin and Cynthia Lake. Effective programs in elementary mathematics: A best-evidence synthesis. *Review of Educational Research*, 78(3):427–515, 9 2008.

[272] Robert E. Slavin, Cynthia Lake, and Cynthia Groff. Effective programs in middle and high school mathematics: A best-evidence synthesis. *Review of Educational Research*, 79(2):839–911, 2009.

[273] J. Smith, A. A diSessa, and J. Roschelle. Misconceptions reconceived: A constructivist analysis of knowledge in transition. *Journal of the Learning Sciences*, 3(2):115–163, 1993.

[274] M. Smith and M. Stein. *5 Practices for Orchestrating Productive Mathematics Discussions. National Council of Teachers of Mathematics*. National Council of Teachers of Mathematics, Reston, VA, 2011.

[275] Daniel Solórzano, Miguel Ceja, and Tara Yosso. Critical race theory, racial microaggressions, and campus racial climate: The experiences of african american college students. *The Journal of Negro Education*, 69(1/2):60–73, 2000.

[276] H. Soto. Learning mathematics through embodied activities, 2016. [Online; accessed 2022-06-01].

[277] Steven J. Spencer, Claude M. Steele, and Diane M. Quinn. Stereotype threat and women's math performance. *Journal of Experimental Social Psychology*, 35(1):4–28, 1 1999.

[278] C. Staats, K. Capatosto, L. Tenney, and S. Mamo. State of the science: Implicit bias review 2017. Technical report, Columbus, OH, 2017.

[279] D. Stables. Asia's isle of five separate genders, 4 2021.

[280] Claude M. Steele. A threat in the air: How stereotypes shape intellectual identity and performance. *American Psychologist*, 52(6):613–629, 1997.

[281] L. A Steen. *Calculus for a New Century: A Pump, Not a Filter: National Colloquium: Papers*. Mathematical Association of America, 1988. (US).

[282] Mary Kay Stein and Margaret Schwan Smith. Mathematical tasks as a framework for reflection: From research to practice. *Mathematics teaching in the middle school*, 3(4):268–75, 1998.

[283] M. Steiner. Mathematical explanation. *Philosophical Studies*, 34(2):135–151, 1978.

[284] A. A. Stewart and E. Swanson. Turn and talk: An evidence-based practice teacher's guide. Technical report, 2019.

[285] David W. Stinson. Negotiating the "white male math myth": African american male students and success in school mathematics. *Journal for Research in Mathematics Education*, 44(1):69–99, 1 2013. publisher: National Council of Teachers of Mathematics section: Journal for Research in Mathematics Education.

[286] Alan Stoskopf. Echoes of a forgotten past: Eugenics, testing, and education reform. volume 66, page 126–133. Taylor & Francis, 2002.

[287] Francis Edward Su. Mathematical microaggressions. *MAA Focus*, pages 36–37, 2015.

[288] Derald Wing Sue, Sarah Alsaidi, Michael Awad, Elizabeth Glaeser, Cassandra Z. Calle, and Narolyn Mendez. Disarming racial microaggressions: Microintervention strategies for targets, white allies, and bystanders. *The American psychologist*, 2019.

[289] Derald Wing Sue, Christina M. Capodilupo, Gina C. Torino, Jennifer M. Bucceri, Aisha Holder, Kevin L. Nadal, and Marta Esquilin. Racial microaggressions in everyday life: implications for clinical practice. *American psychologist*, 62(4):271, 2007.

[290] J. Sunderland. *Gendered discourses*. Palgrave Macmillan, Basingstoke, 2004.

[291] J. H. Tai, Mollie Dollinger, Rola Ajjawi, Trina Jorre de St Jorre, Shannon Krattli, Danni McCarthy, and Daniella Prezioso. Designing assessment for inclusion: an exploration of diverse students' assessment experiences. *Assessment & Evaluation in Higher Education*, 0(0):1–15, 6 2022.

[292] Joanna Tai, Rola Ajjawi, David Boud, Phillip Dawson, and Ernesto Panadero. Developing evaluative judgement: enabling students to make decisions about the quality of work. *Higher Education*, 76(3):467–481, 9 2018.

[293] R. Talbot. Specifications grading: We may have a winner, 4 2017. [Online; accessed 2022-06-13].

[294] R. Talbot. The emrn rubric, 4 2022. [Online; accessed 2022-06-13].

[295] Julie Stulac Tammivaara. The effects of task structure on beliefs about competence and participation in small groups. *Sociology of Education*, 55(4):212–222, 1982. publisher: [Sage Publications, Inc., American Sociological Association].

[296] H. Thiry, T. J. Weston, R. P. Harper, D. G. Holland, A. K. Koch, B. M. Drake, Anne-Barrie Hunter, and Elaine Seymour. *Talking about leaving revisited: Persistence, relocation, and loss in undergraduate STEM education*. Springer, Cham, Switzerland, 2019.

[297] Dustin B. Thoman, Jessica A. Arizaga, Jessi L. Smith, Tyler S. Story, and Gretchen Soncuya. The grass is greener in non-science, technology, engineering, and math classes: Examining the role of competing belonging to undergraduate women's vulnerability to being pulled away from science. *Psychology of Women Quarterly*, 38(2):246–258, 6 2014. publisher: SAGE Publications Inc.

[298] Carla J. Thompson. Educational statistics authentic learning capsules: Community action projects for students utilizing leadership and e-based statistics. *Journal of Statistics Education*, 17(1):null, 3 2009.

[299] Jessica To, Ernesto Panadero, and David Carless. A systematic review of the educational uses and effects of exemplars. *Assessment & Evaluation in Higher Education*, 0(0):1–16, 12 2021.

[300] Kenneth Tobin. Effects of teacher wait time on discourse characteristics in mathematics and language arts classes. *American Educational Research Journal*, 23(2):191–200, 6 1986. publisher: American Educational Research Association.

[301] Kenneth Tobin. The role of wait time in higher cognitive level learning. *Review of Educational Research*, 57(1):69–95, 3 1987. publisher: American Educational Research Association.

[302] Keith Topping. Peer assessment between students in colleges and universities. *Review of Educational Research*, 68(3):249–276, 1998.

[303] Arielle True-Funk, Cristina Poleacovschi, Gloria Jones-Johnson, Scott Feinstein, Kalynda Smith, and Stephanie Luster-Teasley. Intersectional engineers: Diversity of gender and race microaggressions and their effects in engineering education. *Journal of Management in Engineering*, 37(3):04021002, 5 2021. publisher: American Society of Civil Engineers.

[304] Alex Shevrin Venet. *Equity-Centered Trauma-Informed Education*. W. W. Norton & Company, New York, 5 2021.

[305] Susanne Vogel and Lars Schwabe. Learning and memory under stress: implications for the classroom. *npj Science of Learning*, 1(1):1–10, 6 2016.

[306] L. S. Vygotsky. *Mind in society: The development of higher mental process*. Harvard University Press, Cambridge, MA, 1978.

[307] J. Wai. Why is it socially acceptable to be bad at math?, 2012. [Online; accessed 2019-12-02].

[308] Margaret Walshaw and Glenda Anthony. The teacher's role in classroom discourse: A review of recent research into mathematics classrooms. *Review of Educational Research*, 78(3):516–551, 9 2008.

[309] Song Wang, Rou-Jia Sung, Daniel L. Reinholz, and Thomas J. Bussey. Equity analysis of an augmented reality-mediated group activity in a college biochemistry classroom. *Journal of Research in Science Teaching*, n/a(n/a), 2023. _eprint: https://onlinelibrary.wiley.com/doi/pdf/10.1002/tea.21847.

[310] Hiroko Kawaguchi Warshauer. Productive struggle in middle school mathematics classrooms. *Journal of Mathematics Teacher Education*, 18(4):375–400, 8 2015.

[311] Ronald L. Wasserstein and Nicole A. Lazar. The ASA Statement on p-Values: Context, Process, and Purpose. *The American Statistician*, 70(2):129–133, April 2016. Publisher: Taylor & Francis _eprint: https://doi.org/10.1080/00031305.2016.1154108.

[312] Noreen M. Webb, Megan L. Franke, Nicholas C. Johnson, Marsha Ing, and Joy Zimmerman. Learning through explaining and engaging with others' mathematical ideas. *Mathematical Thinking and Learning*, 0(0):1–27, 10 2021. publisher: Routledge.

[313] J. Wellington. Evaluating a mathematics program for adoption: Connecting Math Concepts. *Effective School Practices*, 13(2):70–75, 1994.

[314] Grant Wiggins. The case for authentic assessment. *Practical Assessment, Research, and Evaluation*, 2(1), 11 2019.

[315] Jonee Wilson, Mahtab Nazemi, Kara Jackson, and Anne Garrison Wilhelm. Investigating Teaching in Conceptually Oriented Mathematics Classrooms Characterized by African American Student Success. *Journal for Research in Mathematics Education*, 50(4):362–400, 2019. Publisher: National Council of Teachers of Mathematics.

[316] Karen Wilson and James H. Korn. Attention during lectures: Beyond ten minutes. *Teaching of Psychology*, 34(2):85–89, 4 2007. publisher: SAGE Publications Inc.

[317] Melody Wiseheart, Carolina E. Küpper-Tetzel, Tina Weston, Alice S. N. Kim, Irina V. Kapler, and Vanessa Foot-Seymour. Enhancing the quality of student learning using distributed practice. In *The Cambridge handbook of cognition and education*, pages 550–583. Cambridge University Press, New York, NY, US, 2019. DOI: 10.1017/9781108235631.

[318] Benedikt Wisniewski, Klaus Zierer, and John Hattie. The power of feedback revisited: A meta-analysis of educational feedback research. *Frontiers in Psychology*, 10, 2020. [Online; accessed 2022-09-06].

[319] Paul L. Witt, Lawrence R. Wheeless, and Mike Allen. A meta-analytical review of the relationship between teacher immediacy and student learning. *Communication Monographs*, 71(2):184–207, 6 2004.

[320] Shelley Yijung Wu and Dan Battey. The cultural production of racial narratives about asian americans in mathematics. *Journal for Research in Mathematics Education*, 52(5):581–614, 11 2021. publisher: National Council of Teachers of Mathematics section: Journal for Research in Mathematics Education.

[321] Wuri Wuryandani and Herwin. The effect of the think-pair-share model on learning outcomes of civics in elementary school students. *Cypriot Journal of Educational Sciences*, 16(2):627–640, 2021. publisher: Birlesik Dunya Yenilik Arastirma ve Yayincilik Merkezi.

[322] Veronica X. Yan, Yue Yu, Michael A. Garcia, and Robert A. Bjork. Why does guessing incorrectly enhance, rather than impair, retention? *Memory & Cognition*, 42(8):1373–1383, 11 2014.

[323] Zi Yan and Gavin T.L. Brown. A cyclical self-assessment process: towards a model of how students engage in self-assessment. *Assessment & Evaluation in Higher Education*, 42(8):1247–1262, 11 2017.

[324] David S. Yeager and Carol S. Dweck. What can be learned from growth mindset controversies? *American Psychologist*, 75:1269–1284, 2020. publisher-place: US publisher: American Psychological Association.

[325] David Scott Yeager, Valerie Purdie-Vaughns, Julio Garcia, Nancy Apfel, Patti Brzustoski, Allison Master, William T. Hessert, Matthew E. Williams, and Geoffrey L. Cohen. Breaking the cycle of mistrust: Wise interventions to provide critical feedback across the racial divide. *Journal of Experimental Psychology: General*, 143:804–824, 2014. Place: US Publisher: American Psychological Association.

[326] William Zahner, Ernesto Daniel Calleros, and Kevin Pelaez. Designing learning environments to promote academic literacy in mathematics in multilingual secondary mathematics classrooms. *ZDM – Mathematics Education*, 53(2):359–373, May 2021.

D

Glossary

Academically productive talk. Classroom discourse that promotes student learning and understanding, and encourages students to engage in academic discussions. This higher-level academic discourse that helps students recognize, build on, and respond to the thinking of the teacher and their peers.

Access. Access means that everyone can exist in a space in their full humanity and have their varied needs respected and met. Ensuring access is both a matter of accessible design (i.e., flexible practices that generally work for a lot of different people), and specific accommodations for individual needs that are not adequately addressed by your overall design.

Access needs. The things that an individual needs to access a space. Access needs are many and can be varied, including. 1. Physical access and ergonomics 2. Opportunities to move, stretch, and take breaks 3. Food, drinks, appropriate bathrooms 4. Language/communication that is understandable 5. Safety, trust, and nonjudgment.

Active learning. A phrase used primarily in higher education, which refers to classrooms in which students engage in a variety of activities (as compared to pure lecture).

Assigning competence. Assigning competence is the practice of explicitly elevating the status of students perceived as low status by highlighting their meaningful contributions in public spaces.

Average Contributions. A metric that accounts for the size of various demographic groups when looking at equity data. We calculate this by dividing the overall number of contributions for each group by the number of students in the group. As a result, we can understand student contributions with respect to the demographics of the class. For example, if there were 8 contributions from 4 Black students, this would mean the average contributions were 2 for Black students (i.e., on average, Black students contributed twice). In contrast, the 16 contributions from 20 Latinx students would mean the average contributions for Latinx students were only 0.8 (i.e., on average, Latinx students contributed 0.8 times, or less than once).

Behaviorism. A theory of learning that emphasizes the role of reinforcement and punishment in shaping behavior.

Bodymind. A term that acknowledges that human experience is embodied, recognizing that bodies and minds exist together, and our bodies influence the way that we experience the world.

Cohen's d. A statistical measure of the difference between two groups (i.e., effect size), calculated by dividing the difference between the means of the groups by the standard deviation of the data.

Complex Instruction. Complex Instruction is a set of instructional practices grounded in sociological theory that focuses on the concept of status. High status students tend to have more opportunities to participate, and their ideas are seen as more valuable. Consequently, targeting status as a site of intervention can be a useful strategy for remedying racial (and other) inequities in math classrooms. Complex Instruction uses four primary strategies to reduce status imbalances and promote student belonging (i.e., status interventions).

Constructivism. A theory of learning that emphasizes the active construction of knowledge by learners, through their experiences and interactions with the world. A key tenet of constructivism is that students are not blank slates, but that all new knowledge must be built on prior knowledge.

Direct instruction. A teacher-centered approach to instruction that emphasizes teacher-led lectures, drills, and practice activities (i.e., lecture).

Direction (of feedback). The direction of feedback focuses on whether it aims to highlight something positive (strengths) or or negative (areas of growth).

Disability justice. Disability justice contrasts a medicalized approach to disability, which fixates on individual flaws and accommodations. Instead, disability justice situates individual embodied experiences within broader social structures. Disability justice attends to the pernicious role of ableism in society, especially as it undergirds and amplifies other systems of oppression. A disability justice approach is inherently intersectional, collaborative, and liberatory, insofar that it focuses on dismantling all systems that create hierarchy rather than simply replacing old hierarchies with new ones. This framework was developed in activist spaces—not the academy—through the liberatory work of disabled people of color and disabled queer activists. Disability justice attends explicitly to access, wholeness, embodiment, and humanity in the quest for liberation. It aims to disrupt ideas of normal, normative, and perfect, recognizing the beauty of the imperfect.

D/discourse. The use of lowercase and capital D distinguishes the concept of Discourse in postructuralism from the term discourse, which often is used to refer to classroom discussion. Broadly speaking, a Discourse is a collection of symbols, signs, artifacts, and other cultural representations that work together to constitute the social world. Discourses constrain how individuals act, by creating a limited set of subject positions that they are permitted to occupy. In this way, Discourses are mechanisms of power, because they constrain actions within the social world.

Effect size. A statistical measure of the magnitude of an effect or relationship, calculated by dividing the difference between the means of two groups by the standard deviation of the data.

EQUIP. EQUIP is a customizable observation tool for tracking patterns in student participation. The goal is simple. to empower teachers in building more equitable classrooms. EQUIP can be used in real-time or with videos of classroom teaching. After completing an observation, EQUIP generates instant analytics that teachers can use to improve their practice.

Equity Analytics. The theoretical underpinning of the EQUIP tool, which focuses on creating a fair distribution of resources.

Feedback. Feedback is a process in which learners make sense of information about their performance and use it to enhance the quality of their work or learning strategies.

Group worthy. Group Worthy Tasks are used to engage students in deep learning with problems that require multiple minds to meaningfully engage with.

Hedge's g. An effect size that is essentially the same as Cohen's d but it uses a pooled weighted standard deviation for the populations, which is useful when the samples are significantly different in size.

Implicit bias. Implicit bias refers to unconscious attitudes or stereotypes that affect our understanding, actions, and decisions towards individuals or groups. These biases are usually based on social categorizations such as race, gender, or ethnicity and can impact behavior even when someone consciously rejects the biased belief.

Imposter syndrome. Imposter syndrome is a psychological pattern in which an individual doubts their accomplishments and has a persistent fear of being exposed as a fraud, despite evidence of their competence. People with imposter syndrome often attribute their successes to luck or external factors and feel undeserving of their accomplishments.

Inquiry-based learning. Inquiry-based learning is an approach to education that focuses on promoting curiosity, critical thinking, and problem-solving skills by engaging learners in asking questions and exploring topics through investigation, research, and experimentation. This is a specific paradigm for active learning.

Instruction. Instruction refers to the teaching and learning process where knowledge and skills are transmitted from the teacher to the learner. Effective instruction involves using a variety of techniques and strategies to engage learners, facilitate their understanding, and promote their achievement.

Microaggression. Microaggressions are subtle, often unintentional, behaviors or remarks that communicate negative attitudes or stereotypes towards marginalized groups. These actions can be damaging to individuals and groups and contribute to a hostile environment.

Practice. Practice often refers to the act of engaging in a task or activity repeatedly to improve skills or proficiency. Practice can also refer to the routines, habits, or behaviors that experts engage in within their domain of expertise.

Social Marker Specificity. An underlying principle of equity analytics. This principle states that in order for data to be useful for reducing inequities, the data must be disaggregated along social marker identities (i.e., in order to improve racial equity for Black students, data collected should speak directly to the experiences of Black students, not just equity in general).

Sociocultural Theory. Sociocultural theory is a framework that emphasizes the role of cultural, social, and historical contexts in shaping human development and learning. This theory suggests that people's understanding of the world is socially constructed and that learning occurs through interaction with others in cultural and linguistic communities.

Status. In the context of education, status refers to a person's perceived position or standing within a classroom or academic setting. Status can be influenced by factors such as academic achievement, popularity, or social identity and can impact an individual's experiences and opportunities in the classroom.

Stereotype. A stereotype is a generalized belief or perception about a particular group of people, often based on oversimplified or inaccurate assumptions. Stereotypes can be harmful and perpetuate discrimination and bias.

Stereotype Threat. Stereotype threat is a phenomenon in which individuals feel at risk of confirming negative stereotypes about their social group, which can lead to anxiety, reduced performance, and lower self-esteem.

Teacher immediacy. Teacher immediacy refers to the degree to which a teacher communicates warmth, approachability, and involvement with their students. This can be expressed through nonverbal behaviors, such as eye contact and gestures, and verbal behaviors, such as using humor or providing feedback.

Teacher Press. Teacher press refers to the level and intensity of demands, expectations, and feedback that a teacher provides to their students. This can influence student motivation, engagement, and achievement.